T0269492

Strategic alignment of innovation to business

Strategic alignment of innovation to business

Balancing exploration and exploitation in short and long life cycle industries

Frances Fortuin

Innovation and sustainability series – Volume 2

ISBN 978-90-8686-056-2
ISSN 1874-7663

First published, 2007

Wageningen Academic Publishers
The Netherlands, 2007

Innovation and Sustainability series

The fields of innovation and sustainability are more and more recognized as the major drivers of business success in the 21st century. Today's companies are facing ever-faster changes in their business environment, to which they must respond through continuous innovation. The growing concern regarding the quality and environmental friendliness of products and processes call for fundamentally new ways of developing, producing and marketing of products. New ways of organizing supply chains, with new network ties between firms are needed to cope with these new demands. This series aims to assist industry to conduct the (interorganizational) innovations needed to meet the challenges that are fundamental for the transition from a production orientation to a 'cradle-to-cradle' demand-orientation. However, innovation can be disruptive, not only concerning the organization of the processes, but also regarding the allocation of resources and power bases. Existing companies are increasingly challenged by newcomers, e.g. start-up firms and spin-off ventures. In the transition process, supplier bases might be reorganized, activities reallocated, and relations and role allocations changed as new entities occur. We want to study these new organizational forms and their consequences – as we view them as core for these business networks in transition.

About the Editor

Onno Omta is chaired professor in Business Administration at Wageningen University and Research Centre, the Netherlands. He received an MSc in Biochemistry and a PhD in innovation management, both from the University of Groningen. He is the Editor-in-Chief of The Journal on Chain and Network Science, and he has published numerous articles in leading scientific journals in the field of chains and networks and innovation. He has worked as a consultant and researcher for a large variety of (multinational) technology-based prospector companies within the agri-food industry (e.g. Unilever, VION, Bonduelle, Campina, Friesland Foods, FloraHolland) and in other industries (e.g. SKF, Airbus, Erickson, Exxon, Hilti and Philips).

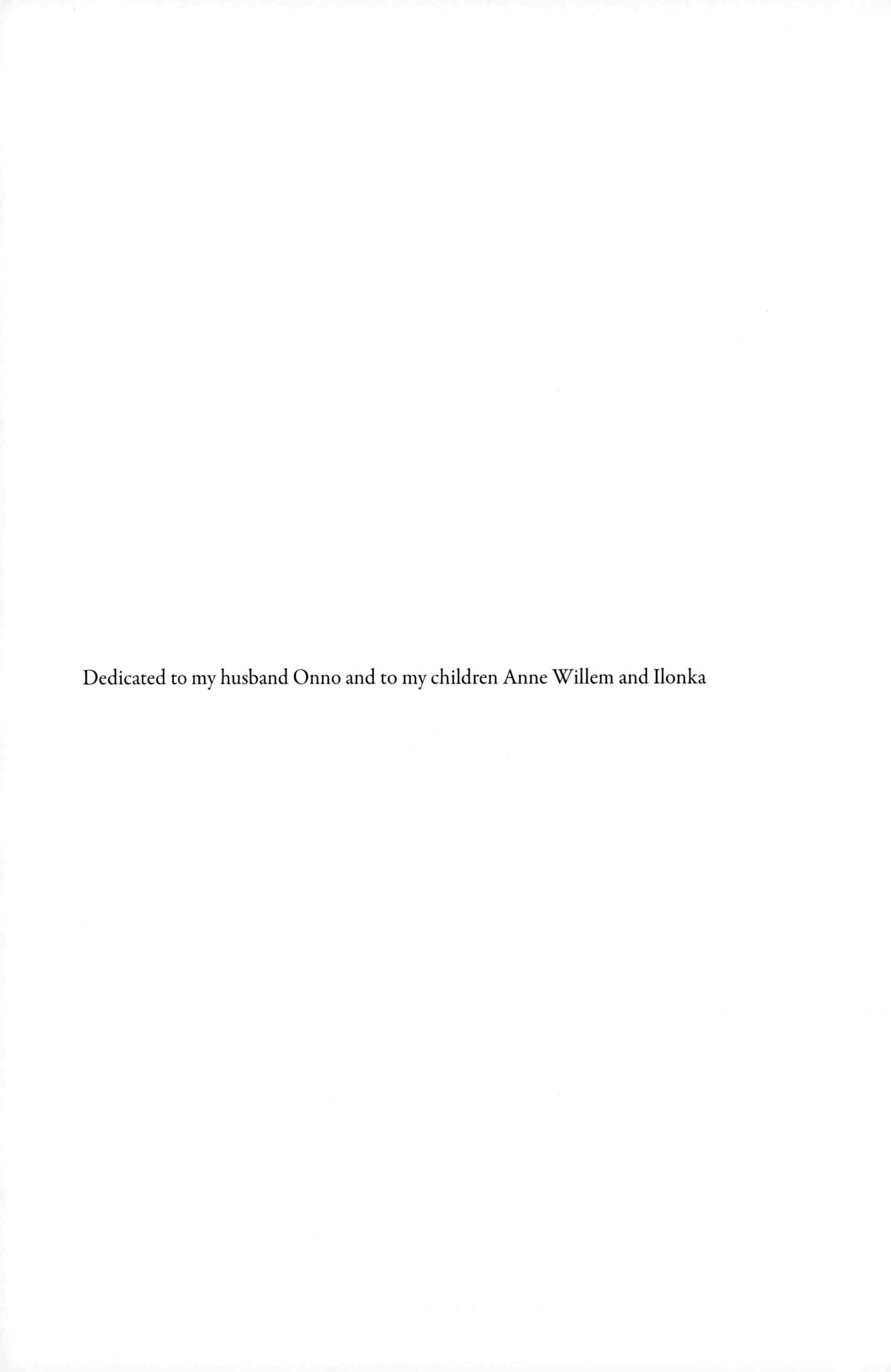

Dedicated to my husband Onno and to my children Anne Willem and Ilonka

Table of contents

Preface 9

1. Introduction 11

2. Theoretical perspectives on strategy 15
2.1 Strategic management 15
2.2 The history of strategic management thought 17
2.3 Static versus dynamic strategy models 18
2.4 The two main theoretical perspectives on strategy 19
2.5 The industrial organization perspective 19
2.6 Competence perspective 22
2.7 The concept of strategic alignment 27
2.8 Concluding remarks 30

3. Study domain: innovation 31
3.1 The phenomenon of innovation 31
3.2 Innovation typologies 32
3.3 The R&D **process** **34**
3.4 Barriers to innovation 37
3.5 Drivers of innovation 38
3.6 Best practices in innovation 42
3.7 Innovation strategy and the two perspectives on strategy 45
3.8 Analyzing strategic alignment in the case of innovation 47
3.9 Concluding remarks 49

4. Research design 51
4.1 The general conceptual framework 51
4.2 Overall research design 52
4.3 The cross-industry study 54
4.4 The longitudinal study 65
4.5 Concluding remarks 74

5. Cross-industry study results 75
5.1 Data collection and baseline description of the cross-industry companies 75
5.2 R&D Performance 78
5.3 Strategic alignment 81
5.4 R&D strategy 83
5.5 R&D competencies 86
5.6 R&D capabilities 89
5.7 R&D process 97

5.8 Methods to improve strategic alignment 101
5.9 Concluding remarks 104

6. Longitudinal study results **107**
6.1 Baseline description 107
6.2 Study sample, reliability and validity 108
6.3 Comparison of R&D and BU/headquarters assessments 109
6.4 Factors related to strategic alignment (external fit) 111
6.5 The longitudinal analyses 113
6.6 Concluding remarks 122

7. Discussion and conclusions **123**
7.1 The cross-industry study 124
7.2 The longitudinal study 128
7.3 General conclusion 131
7.4 Implications for innovation management 133
7.5 Concluding remarks 136

References **139**

Appendices **153**
Appendix A. General questions in the cross-industry study 153
Appendix B. Cross-industry survey questionnaire I 155
Appendix C. Cross-industry survey questionnaire II 163
Appendix D. Longitudinal survey questionnaire 166
Appendix E. terms used in the research questionnaires 168

About the author **171**

Preface

For the several employments and offices of our fellows, we have twelve that sail into foreign countries under the names of other nations (for our own conceal), who bring us the books and abstracts, and patterns of experiments of all other parts. These we call Merchants of Light. We have three that collect the experiments which are in all books. These we call Depredators. We have three that collect the experiments of all mechanical arts, and also of liberal sciences, and also of practices which are not brought into arts. These we call Mysterymen. We have three that try new experiments, such as themselves think good. These we call Pioneers or Miners. We have three that draw the experiments of the former four into titles and tables, to give the better light for the drawing of observations and axioms cut of them. These we call Compilers. We have three that bend themselves, looking into the experiments of their fellows, and cast about how to draw out of them things of use and practice for man's life and knowledge, as well for works as for plain demonstration of causes, means of natural divinations and the easy and clear discovery of the virtues and parts of the bodies. These we call Dowrymen or Benefactors. Then after divers meetings and consults of our whole number to consider of the former labors and collections, we have three that take care out of them to direct new experiments, of a higher light, more penetrating into Nature than the former. These we call Lamps. We have three others that do execute the experiments so directed and report them. These we call Inoculators. Lastly, we have three that raise the former discoveries by experiments into greater observations, axioms, and aphorisms. These we call Interpretators of Nature. We have also, as you must think, novices and apprentices, that the succession of the former employed men do not fail; beside a great number of servants and attendants, men and women. And this we do also: We have consultations, which of the inventions and experiences which we have discovered shall be published, and which not: and take all an oath of secrecy for the concealing of those which we think fit to keep secret: though, some of those we do reveal sometimes to the State, and some not.

Nova Atlantis (Sir Francis Bacon (1625) in Omta, 1995).

About 1625, Sir Francis Bacon magnificently described an organization that was totally innovation oriented in the ideal world of Nova Atlantis. Today, R&D (Research and Development) organizations can be found in the everyday world. This book will bring you into this world. It describes the day-to-day activities of innovation management and the constant struggle to align innovation to business strategy. It aims to answer such questions as:

- What causes the (lack of) strategic alignment between innovation and business?
- How can strategic alignment be achieved and maintained?

This book presents the findings of a cross-industry study into the management and organization of innovation in ten technology-based companies in different industries typified by the length of their product generation life cycles (PGLCs) and a six-year longitudinal study in one of these companies to research the dynamics of strategic alignment. It aims to provide a sound empirical basis for a number of ideas and statements about innovation management in general. Parts of the book have been presented at scientific congresses and workshops, and the findings have served as a platform for discussion with managers in R&D. The study was undertaken at the Department of Business Administration at Wageningen University.

I would like to thank the CTOs and R&D Directors of the companies that participated in the cross-industry study, who gave up precious time for interviews about innovation management, and whose insights greatly enhanced my understanding of the subject of this book. In addition, my thanks go to the R&D department heads and R&D program managers of these companies for their willingness to fill out the self-assessment questionnaires. I am very grateful to the company that provided me with a unique opportunity to conduct a longitudinal study. Without the participation of the R&D staff and higher management of the strategic business units and corporate headquarters in completing the research questionnaires up to four times, this study would not have been possible.

Frances Fortuin, September 2007.

1. Introduction

The field of innovation is widely recognized as one of the major drivers of business success in the 21st century. The American Management Association (AMA) concluded, based on a survey of 1,396 top executives in large multinational companies, that more than 90% of them consider innovation to be (extremely) important for their company's long-term survival, with over 95% considering that this will still be the case in ten years' time (Jamrog, 2006). However, identifying innovation as vital for business is not enough. A recent Booz Allen & Hamilton survey found that just spending more on R&D (Research and Development) does not necessarily equate with greater innovation outcomes (Jaruzelski *et al.*, 2005). Indeed, Edler *et al.* (2002) conclude from a survey of more than 200 US, Japanese and European companies with annual R&D budgets of over US$ 100 m that only if innovation is adequately linked to corporate strategy it really pays off in terms of accelerated corporate sales growth. Unfortunately, the 2005 AMA Innovation Survey revealed that 85% of the executives did not consider their firms to be very successful at executing innovation strategy. As Clark (2006) indicates, it is not the lack of a strategy that causes a business to fail but rather the firm's inability to act upon a chosen strategy. This raises the question why it is so hard to turn strategic intent into action in the field of innovation.

The answer lays in the important challenge for companies that want to pursue an innovation strategy, namely to bridge the gap between exploration (searching for new knowledge) and exploitation (exploiting existing knowledge). The distinction between exploration and exploitation goes back to Holland (1975) and was further developed by March (1991). The long-term orientation needed for exploration, and the high inherent uncertainty of its outcomes is regarded being at odds with the predictability needed for executing the day-to-day activities efficiently (e.g. Roberts, 1995; Park and Gil, 2006). Exploration is not about efficiency of current activities, it is an uncertain process that deals with the search for new opportunities Kline (1986) rightly specifies (radical) innovation as inherently disorderly: *'Models that depict innovation as a smooth, well-behaved linear process badly mis-specify the nature and direction of the causal factors at work. Innovation is complex, uncertain, somewhat disorderly, and subject to changes of many sorts.'* In contrast to exploration, exploitation adds to the existing competencies and capabilities of the firm without changing the nature of these activities. Business unit managers prefer R&D to come up with exploitative innovations, incrementally moving the performance bar a little bit higher, without infringing upon their complex set of technological and business relationships (Anderson and Tushman, 1990). However, for the long term survival of their firms, top management has to balance exploration and exploitation in order to withstand the constant threat of new entrants and technological change in today's highly dynamic business environment. In order to do so, companies are increasingly perforating the boundaries of their firms, e.g. by starting alliances with (start-up) firms, and building up internal venture groups scouting for new ideas, products and processes outside the firm. This recent transition to more 'open' forms of innovation (Chesbrough,

2003) has made the management task of the strategic alignment of innovation to business even more compelling.

Although general consensus exists among strategy scholars about the importance of strategic alignment (e.g. Porter, 1985; Shrivastava *et al.*, 1992; Mintzberg and Lampbel, 1999; Johnson and Scholes, 2002) in general and in innovation in particular (e.g. Brockhoff *et al.*, 1999; Tidd *et al.*, 2001; Storey, 2004; Fagerberg *et al.*, 2004), the factors and mechanisms that underlie the process of achieving and maintaining strategic alignment are much less explored. In approaching this issue we build upon one of the most widely shared and enduring assumptions in the strategy literature, which postulates that, if it is to be effective, a strategy has to be in accordance with the external as well as the internal contingencies of a firm, also referred to as the external and internal fit (e.g. Burns and Stalker, 1961; Lawrence and Lorsch, 1967; Ginsburg and Venkatraman, 1985; Venkatraman, 1989; Miles and Snow, 1994; Burton and Obel, 1998; Verdú Jover *et al.*, 2005; Katsikeas *et al.*, 2006). Where internal fit requires that a chosen strategy is in compliance with the firm's internal structures and processes, external fit demands that a firm matches its strategy with the opportunities and threats provided by the external environment (Lawrence and Lorsch, 1967; Thompson, 1967). Strategic alignment is thus concerned with finding the right balance between the relevant contingencies in the business environment (external fit) and the firm's internal resources, competencies and capabilities (internal fit). The process of strategic alignment is inherently dynamic because strategic choices made by a firm will inevitably evoke counteractions (e.g. imitation, own innovations) by its major competitors, which will necessitate a subsequent response. Strategic alignment is, therefore, not an event but a process of continuous adaptation and change (Ginsberg, 1988), Henderson and Venkatraman, 1993). Understanding strategic alignment of innovation to business thus means that we have to investigate which factors determine the alignment of a firm's innovation strategy with its external environment as well as its internal resources, competencies and capabilities.

Two interrelated studies were conducted to address the main question posed in this book.

How can technology-based firms achieve strategic alignment of innovation to business?

The first study explores the factors that affect strategic alignment of innovation to business across industries. We pose that it is possible to compare among industries, provided they are classified according to the industry 'clockspeed' (Brown and Eisenhardt, 1998), indicated by the length between the subsequent product generations, further referred to as the Product Generation Life Cycle (PGLC). The empirical data were collected in a cross-industry survey including ten large, multinational technology-based companies, world leaders in their respective industries. The average PGLCs in these industries range from just several months in electronics and the mobile phone industry to (more than) 10 years in aerospace and pharmaceutics. The research question that is addressed by this study is the following:

RQ1. What is the effect of the industry 'clockspeed' on the strategic alignment of innovation to business?

The fact that technology-based companies typically operate under highly dynamic market and technology conditions forces them to continually adapt their competencies and capabilities to the rapidly changing business environment. To study these dynamic aspects of strategic alignment, in one of the ten companies -a large multinational supplier of industrial components- a six-year biannual longitudinal study was conducted covering the R&D center, the corporate headquarters and the business units. The research question that is addressed by this study is the following:

RQ2. How can strategic alignment of innovation to business be achieved and maintained over time?

Chapter 2 discusses the two main theoretical perspectives used to understand how a firm can gain and maintain competitive advantage, namely the industrial organization perspective (also referred to as the outside-in approach) and the competence perspective (also referred to as the inside-out approach). In the industrial organization perspective, the focus of analysis is external. A major concern here is how the firm compares to its industry competitors by emphasizing the actions a firm can take to create a defensible position against competitors (Porter, 1980, 1998). This approach views the essence of competitive strategy formulation as relating a firm to its business environment. It implies that the industry structure, as approximated by the length of the PGLC in the cross-industry survey, strongly influences the strategies potentially available to firms. In contrast to this, the competence perspective takes the firm's own resources (including the firm's financial, physical and organizational assets), competencies (skills and knowledge) and capabilities (management systems) as the starting point for gaining competitive advantage. Grant (1996) suggests that, in dynamically-competitive markets, gaining and keeping competitive advantage is more likely to be associated with resource and capability-based advantages than with positioning advantages resulting from market and segment selection and the firm's competitive position within the industry structure. He furthermore reasons that such resource and capability-based advantages are likely to derive from superior access to and the integration of specialized knowledge. We argue that the industrial organization theory and the competence perspective are complementary to understanding internal and external fit and, ultimately, the phenomenon of strategic alignment.

In Chapter 3 the concept of innovation is first introduced, followed by the process of crafting and implementing an innovation strategy. Here the innovation resources, competencies and capabilities, as well as the processes needed to implement an innovation strategy are described. We consider whether the R&D function is a key function in developing new products, processes and services in technology-based firms. Finally, the concepts of innovation and alignment are confronted with the two main perspectives on strategic management, the industrial organization theory and the competence perspective.

In Chapter 4 the research design is discussed. The conceptual framework and the main propositions regarding the phenomenon of strategic alignment that underlie the two empirical studies are elaborated on. The different theoretical elements used in these studies; i.e. the concepts, the observational relationships and the measures taken to provide for complete coverage of the relevant relations and for internal and external validity are also described. Attention is paid to the sampling procedures, the inclusion criteria and the measures taken to ensure the representativeness of the study samples. In addition, the univariate and multivariate methods of data analysis are discussed. The chapter ends with a description of the methods used to approach the study population in both studies.

In Chapter 5 the main results of the cross-industry study are described. On-site visits to the corporate R&D centers of the companies were conducted. Structured interviews were held with the CTOs (Chief Technology Officers), the Directors of the Corporate R&D centers, and the Technology Directors. One questionnaire that contained quantitative and factual information regarding the company as a whole (e.g. sales volume, profitability, and market share of the different BUs), and specific for the corporate R&D center (e.g. R&D budget, R&D personnel, number of patents and R&D management systems) was filled out per company by the CTO or the Director of the Corporate R&D center. Another questionnaire that asked for personal opinions about the quality of the R&D competencies and capabilities, and the level of strategic alignment was presented to the R&D department heads and R&D program managers. The general questions about innovation management posed to the CTOs and the R&D managers and the Research Questionnaires can be found in the Appendices A, B and C, respectively). In addition, we elaborate on the management practices conducted in the different R&D laboratories and investigate their importance for practitioners in innovation management.

The focus of attention shifts in Chapter 6 to the dynamic character of the strategic alignment of innovation to business. A biannual longitudinal survey was conducted from 1997 to 2003 to assess the alignment of the corporate R&D center to its business unit customers, including 696 respondents in total. The Research Questionnaire that was used is presented in Appendix D. Appendix E provides a glossary of the terms used in the research questionnaires, which can also be helpful for readers unfamiliar with the general terminology used in R&D management.

Finally, in Chapter 7 a synopsis of the key findings and their theoretical implications is presented by placing the cross-industry study (Chapter 5) and the longitudinal study (Chapter 6) in a broader theoretical perspective. The roles of the industrial and the competence perspectives are assessed in relation to their respective and combined contributions to an understanding of the phenomenon of aligning innovation to business. The chapter then draws conclusions about the theoretical and methodological contribution of the studies, and the possibilities for further research. The chapter ends with a discussion of the managerial aspects of the strategic alignment of innovation to business.

2. Theoretical perspectives on strategy

The aim of this chapter and Chapter 3 is to position the core issue of this research, the strategic alignment of innovation to business, in a theoretical framework. Section 2.1 introduces the concept of strategic management, the different strategic management layers in an organization, and the different phases within the strategic management process. Section 2.2 provides a short introduction to the history of strategic thought, while Section 2.3 elaborates on static versus dynamic models of strategy making. Section 2.4 discusses the fundamental differences between the two main perspectives on strategy, the industrial organization perspective and the competence perspective. Section 2.5 discusses the industrial organization perspective, while Section 2.6 concentrates on the two major research streams in the competence perspective, the resource-based view and the dynamic capabilities theory. Section 2.7 introduces the concept of strategic alignment as a core dynamic capability for technology-based firms and discusses the different schools of alignment research. This chapter ends with some concluding remarks in Section 2.8.

2.1 Strategic management

Following Omta and Folstar (2005), strategy is defined as the long-term orientation of an organization, and more precisely as:

A series of goal-directed decisions and actions that match an organization's skills and resources with the opportunities and threats in its environment, to meet the needs of markets and to fulfill stakeholder expectations.

Strategic management is then the process of formulating and executing a firm's strategy. It provides the overall direction for the enterprise by specifying the firm's objectives, developing policies and plans to achieve these objectives, and allocating resources to implement these policies and plans. The process involves matching the company's strategic advantages to its business environment, while at the same time a firm's strategy must be executable in the light of its resources, competencies and capabilities. To be effective, corporate strategy should therefore integrate the organizational goals, policies and action sequences (tactics) into a cohesive whole, based on business reality. The reader should bear in mind, however, that although a sense of direction is important, adhering too strictly to strategy can stifle creativity, especially if it is rigidly enforced. In an uncertain and ambiguous world, fluidity can be more important than a finely tuned strategic compass. Companies may fail despite an 'excellent' strategy because the world changes in a way they failed to anticipate. Strategic management is, therefore, basically a dynamic process requiring continuous reassessment and reformation. It involves a complex pattern of actions and reactions and is partially planned and partially emergent, dynamic, and interactive. In multinational firms that serve internationally dispersed markets and produce a wide range of products or services, strategic decisions are likely to be especially complex. They often have to be made in situations of uncertainty and may involve subjective judgments of future developments, about which managers can never be certain.

Firms often summarize their goals and objectives in a mission and/or vision statement. Many people mistake vision for mission, but the two are fundamentally different. A mission statement defines the purpose or broader goal for being in existence or in business. It serves as a guide in times of uncertainty that the mission can remain the same for decades if crafted correctly. A vision statement, by contrast, describes where the goal-setters want to see themselves in the future. It may describe how they see events unfolding in the 10 to 20 years to come if everything goes exactly as they hope. A vision is specific in terms of objectives and time frames of achievement. For example, to help transport goods and people efficiently without damaging the environment is a mission statement, but we will be among the top three transporters of goods and people in Europe by 2010 is a vision statement. Features of an effective vision statement include clarity and un-ambiguity, achievable aspirations and realistic time horizons.

In most large firms, there are several strategic layers. While corporate strategy sets the firm's overall direction and is concerned with the question of what business(es) the company is in or wants to be in, the functional and business unit strategies indicate how the functional departments and strategic business units will contribute to corporate strategy by indicating how they will compete in their specific business or industry. Functional strategies are made up of the goal-directed decisions and actions of the firm's various functional departments. These departments include, for instance, manufacturing, finance and accounting, marketing, purchasing and R&D. The emphasis is on short- and medium-term plans and is limited to the domain of each department's functional responsibility. Many companies now feel that a functional organization is not an efficient way of organizing activities so they have reorganized into strategic business units. A BU is a semi-autonomous unit within the firm treated as an internal profit centre by corporate headquarters. It is usually responsible for its own budgeting, innovation, hiring, and price setting decisions. Each BU is responsible for developing its own business strategy that has to be in line with the broader corporate strategy.

We identify four phases in the strategic management process: the strategic analysis phase, the strategy formulation phase, the implementation and execution phase, and the strategy evaluation phase. In the *strategic analysis phase* (which includes *scanning* and *idea generation*), signals from the business environment about potential opportunities and threats are detected. The processes in the strategic analysis phase can be facilitated by employees fulfilling 'boundary spanning' and 'gate keeping' roles (Brown and Eisenhardt, 1995; Reid and De Brentani, 2004). The strategic objectives are set in the *strategy formulation phase*. This involves crafting vision statements (the long-term view of a possible future), mission statements (the role that the organization gives itself in society), overall corporate objectives (both financial and strategic), strategic business unit objectives (both financial and strategic), and tactical objectives. These objectives should, in the light of the situation analysis, suggest a strategic plan. The plan provides details of how to achieve these objectives. This three-step strategy formation process is sometimes referred to as determining where you are now, determining where you want to go, and then determining how to get there. The *strategy implementation and execution phase* comprises the process of planning (implementation) and actions (execution), with decision-making taking place at gradually lower levels in the organization. It involves the allocation of sufficient resources (financial, personnel, time, and computer system support), establishing a chain of command or some alternative structure (such as cross-functional teams) and assigning responsibility of specific tasks or processes to specific individuals or groups. In the *evaluation phase,* the information gathered in the previous phases is used to identify possibilities for improvement. This includes monitoring results, comparing benchmarks and best practices, evaluating the efficacy and efficiency of the process, controlling for variances, and making adjustments to the process as necessary. After evaluation, the cycle starts again with a new strategic analysis, using the insights gained in the preceding cycle. This is reflected in the model proposed by Rosenbloom and Burgelman (1989) that sees the process of strategy making as essentially a learning process based on knowledge about which actions have led to past success or failure.

2.2 The history of strategic management thought

Although fundamental work on strategy was carried out in the first half of the 20th century, the main growth of strategy literature took place from the 1950s onwards. We will discuss below the five founders of strategic management thought, Selznick (1957), Chandler (1962), Ansoff (1965), Drucker (1985) and Mintzberg (1999).

Selznick (1957) was the first to introduce the idea of matching an organization's internal factors to its external environmental circumstances. His core idea is that the strengths and weaknesses of a firm are assessed in the light of the opportunities and threats from the business environment. Chandler (1962) recognized the importance of coordinating the various aspects of management under one all-encompassing strategy. Prior to this, the various functions of management have been separated with little overall coordination. Interactions between functions or between departments were typically handled by a boundary position, that is, there were one or two managers who relayed information back and forth between the departments. Chandler also stressed the importance of taking a future-looking long-term perspective. In his groundbreaking work 'Strategy and Structure' (1962), he showed that a long-term coordinated strategy was necessary to give a company structure, direction and focus. 'Structure follows strategy' was his famous phrase.

Ansoff (1965) built on Chandler's work by adding a strategy grid that compared innovation, market penetration, market development, and diversification strategies. In his classic 'Corporate Strategy' (1965) he developed the 'gap analysis' to understand the gap between where a firm currently stands and where it would like to be, to help develop 'gap reducing actions'. Ansoff classified strategic management into three main schools: (1) management by control of performance (after the fact), which is adequate when change is slow; (2) management by projection, when the future can be predicted by extrapolation from the past, and; (3) management by anticipation, when change is slow enough to permit timely anticipation and response to discontinuities. In 1982 he saw the arrival of a new era of rapid, discontinuous change and then identified a fourth strategy: (4) management through flexibility and rapid response to environmental change (Ansoff, 1982).

Drucker was a strategy theorist with a career spanning five decades. His contributions to strategic management were numerous but two are particularly important. Firstly, he stressed the importance of objectives. An organization without clear objectives is like a ship without a rudder. In 1954 he developed the theory of management by objectives, indicating that the procedure of setting objectives and monitoring progress towards them should permeate an entire organization. His other seminal contribution was in predicting the importance of what today we call 'intellectual capital'. He predicted the rise of what he called the 'knowledge worker' and explained the consequences of this for management. He indicated that knowledge work should be carried out in teams with the most knowledgeable person for the task at hand being the temporary leader.

Finally, Mintzberg and Quinn (1991) concluded that the strategy process was much more fluid and unpredictable than people thought. Based on this observation, they defined five types of strategies; strategy as a plan: a direction, guide, course of action, and intention; strategy as a ploy, a maneuver intended to outwit a competitor; strategy as a pattern, a consistent pattern of past behavior, realized rather than intended; strategy as a position, the location of brands, products, or the company; and strategy as a perspective, determined primarily by a master strategist.

2.3 Static versus dynamic strategy models

Several theorists have problems with the static model of strategy: it is not how it is done in real life, for strategy is basically a dynamic and interactive process. Some of the earliest challenges to the planned strategy approach came from Lindblom (1959), who claimed that strategy is a fragmented process of serial and incremental decisions. He viewed strategy as an informal process of mutual adjustment with little apparent coordination. Mintzberg (1979) also made a distinction between deliberate strategy and emergent strategy. Emergent strategy originates not in the mind of the strategist, but in the interaction of the organization with its environment. He claims that emergent strategies tend to exhibit a type of convergence in which ideas and actions from multiple sources integrate into a pattern. This is a form of organizational learning. According to this view, organizational learning is in fact one of the core functions of any business enterprise (see also Peter Senge's 'The Fifth Discipline' 1990). Quinn (1980) elaborated on this by developing an approach of 'logical incrementalism'. He claimed that strategic management involves guiding actions and events towards a conscious strategy in a step-by-step process. With regard to the nature of strategic management he said: *Constantly integrating the simultaneous incremental process of strategy formulation and implementation is the central art of effective strategic management*. Burgelman (1988) took this thought one step further by stating that strategic decisions are not only made incrementally rather than as part of a grand unified vision, but that they are also typically made by numerous people at all levels of the organization. Moncrieff (1999) developed the model of strategy dynamics further. He recognized that strategy is partially deliberate and partially unplanned. The unplanned element comes from two sources, 'emergent strategies' result from the emergence of opportunities and threats in the environment and 'strategies in action' are *ad hoc* actions by people from all parts of the organization. These multitudes of small actions are typically unintentional, informal, and not even recognizable as strategic. In this model, strategy is at the same time planned and emergent, and dynamic and interactive.

2.4 The two main theoretical perspectives on strategy

Currently, the two main approaches to evaluate a firm's strategic position and assess how it can gain and maintain a competitive advantage, are the industrial organization perspective, and the competence perspective (rooted in the evolutionary perspective), encompassing both the resource-based view, and the dynamic capability view (e.g. Truijens, 2004). In the industrial organization perspective, the focus of analysis is external with a major concern being how the firm compares to its industry competitors. It emphasizes the actions a firm can take to create a defensible position against competitors. This approach views the essence of competitive strategy formulation as relating a firm to its business environment. It implies that the industry structure strongly influences the strategies potentially available to firms. Although many studies have adopted the industrial organization perspective, a major question that needs to be addressed is whether strategy is derived entirely from environmental conditions or whether there is a dual relationship between a firm's strategy and its environment. Most classic studies have assumed a 'reactive' perspective, i.e. that strategy needs to be fitted to the environmental conditions; but recent thinking is to attribute a proactive role to strategy. Proponents of the competence perspective take the latter approach by indicating that a firm's resources - including the firm's financial, physical, human, intangible, and organizational assets - are more important than the industry structure in gaining and keeping competitive advantage. Table 2.1 provides an overview of the two perspectives, the founding authors in each theoretical stream and the premises of how competitive advantage is achieved.

2.5 The industrial organization perspective

Industrial organization theorists (e.g. Milgrom and Robberts, 1990; Collis and Montgomery, 1995; Porter, 1998) emphasize the importance of industry forces that provide the opportunities for competitive advantage, defined as a positional advantage derived by a firm which, compared to the competition, provides its customers with lower costs or perceived uniqueness. Two sets of studies are particularly relevant. The first includes studies on strategic groups, especially the Purdue studies (Hatten and Schendel, 1977; Schendel and Patton, 1978), which highlight

Table 2.1. Comparison of the industrial organization and the competence perspectives on strategy. Based on Hunt (2000) and Omta and Folstar (2005).

Theoretical perspective	Premises	Founding authors
Industrial organization perspective	The focus is external. Profitability determinants are industry characteristics and the firm's industry position.	Mason (1939); Bain (1954; 1956); Porter (1980; 1985)
Competence perspective	The focus is internal. Profitability determinants are type, amount and the unique nature of the firm's resources, competencies and capabilities.	
Evolutionary perspective	Competitive dynamics are disequilibrium-provoking. Firms' resources are heterogeneous, path dependency is possible.	Schumpeter (1934; 1942); Alchian (1950); Nelson and Winter (1982); Langlois (1986); Dosi *et al.* (1988); Witt (1992); Hodgson (1993); Foss (1994)
Resource-based view	Resources can be tangible or intangible. Firms are historically situated and resources are heterogeneous and imperfectly mobile.	Penrose (1959); Lippman and Rumelt (1982); Wernerfelt (1984); Dierickx & Cool (1989); Prahalad and Hamel (1990); Barney (1991); Conner (1991); Grant (1991); Hamel and Prahalad (1989; 1994)
Dynamic capabilities framework	Competition is a dynamic disequilibrium-provoking process. Capabilities are dynamic competencies and resources. The continual renewal of the dynamic capabilities stimulates proactive innovation.	Selznick (1957); Andrews (1971); Hofer and Schendel (1978); Teece and Pisano (1994); Day and Nedungadi (1994); Aaker (1995); Heene and Sanchez (1996); Sanchez and Heene (1997); Sanchez (2001)

the need to formulate differential strategies according to the conditions stipulated by the strategic groups and not the entire industry. The second set of studies, especially those by Christensen and Montgomery (1981) and Rumelt (1982), use diversification strategy and market structural variables to explain performance differences. Researchers have argued that competitive advantages should be sustainable to be strategically relevant (Porter, 1980; Coyn, 1985). Sustainable competitive advantage is then defined as a competitive advantage that is not easily replicable or eliminable, that can be maintained over a certain period of time and that is the origin of a firm's sustained superior performance.

Each company is surrounded by other players in its industry environment and by factors over which it has little control. The general environment can be described using PEST factors (Johnson and Scholes, 2002). These include: (1) the political/legal, (2) the economic, (3) sociocultural/demographic, and (4) technological factors. The political/legal factors include: laws, regulations, judicial decisions and political forces at the local, as well as the national and international level. International firms have to know and abide by the national laws and regulations of the countries in which they operate. Management should keep track of changes in each country that could affect their firms in a positive or negative way. The economic factors include macroeconomic data, current statistics and trends, while the sociocultural/ demographic factors encompass the traditions, values, attitudes, beliefs, tastes and patterns of behavior, of the countries in which the company is present. It is important to keep abreast of all relevant changes by following the trends in statistical data, e.g. in population characteristics, to understand current and emerging customer needs. Technological factors indicate what opportunities and threats can be expected from the technological side, e.g. whether or not the firm's products will be affected by rapidly changing technology. The source of this information is usually industry specific.

The industry environment can be characterized by its degree of turbulence, complexity, dynamics and (un-)predictability. The main forces affecting companies in an industry are summed up by Porter's (1985) 'five-major-forces' framework: (1) rivalry among existing firms, (2) the threat of new entrants, (3) the threat of substitute products or services, (4) the bargaining power of suppliers, and (5) the bargaining power of buyers. The interplay of these five forces is thought to determine the boundaries for the firm's competitive strategy. The competitive forces model can help a firm to position itself in an industry in such a way that it can best defend itself or influence the forces at play in its favor. Below we elaborate on the five forces separately.

2.5.1 Rivalry among existing firms

Industry rivalry is likely to be intense if a limited number of companies are striving for dominance. For example, for many years, Coca-Cola was the industry leader and the other players occupied their subordinate positions and accepted their profits. When Pepsi decided to challenge Coke's position of leadership in the 1960s, the industry became intensely and bitterly competitive. All the players were threatened as Coke and Pepsi expanded into every niche of the market by adding new products. Another factor leading to intense competition is a limitation in the possibilities for market expansion because the only way to grow is to take the market share away from competitors. Especially when there is little possibility for differentiation advantages, such as in commodity products where customers can easily switch to a competitor, competition can be brutal.

2.5.2 Threat of new entrants

New competitors can be repelled by several entry barriers. High capital requirements for production, such as in the oil industry, or for R&D, such as in the pharmaceutical industry, may form an effective barrier for new entrants. But brand loyalty too, established by continually advertising the brand and company name, patent protection, high product quality and after-sales service may also make it hard for customers to change to a new, competing product. Absolute cost advantages can act as another entrance barrier: it is hard to compete against a firm with lower costs if their product is of comparable quality. This is one major reason why many US and European companies are moving their plants to India and China. Economies of scale may serve as entrance barriers too: Small businesses thrive on serving market niches that are too small for large firms to serve profitably. This is a special case of differentiation, and one which, in a general sense, can be seen as the ultimate entry barrier. Like lower costs, differentiation can be achieved in virtually any of a company's operations. Finally, legislation can also have a significant effect on entry barriers.

2.5.3 Threat of Substitutes

Some products are direct substitutes for one another: for example, aspartame for sugar. An absence of close substitutes may give a firm the chance to increase prices and profit margins. But newly created substitutes can cancel the advantages a firm has gained.

2.5.4 Power of suppliers

Supplier power is likely to be high when there is a concentration of suppliers rather than a fragmented source of supply, and the costs of switching from one supplier to another are high, e.g. the cost and learning curve associated with a firm changing from one software application to another. It is possible for a supplier to integrate forward if they do not obtain the prices and margins they want in their present business.

2.5.5 Power of buyers

The factors that increase a buyer's power are the mirror image of those that increase a supplier's power. Thus, buyers have enhanced power when they are concentrated and buy in volume, and when there are alternative sources of supply and it costs little to switch between them.

2.6 Competence perspective

The competence perspective relates to the evolutionary economics theoretical stream (Nelson and Winter, 1982) and encompasses the resource-based view (Penrose, 1959) and the dynamic capabilities framework (Teece *et al.*, 1997). Table 2.2 shows an overview of the basic

Table 2.2. Overview of the basic propositions of the competence perspective, the resource-based view and the dynamic capabilities framework. Adapted from Sanchez (2001).

Competence perspective

- Firms have certain 'core' competencies that span products and businesses, change more slowly than products, and arise from collective learning. Firms compete and achieve competitive advantage through creating and using their core competencies.
- Knowledge resources are key sources of competitive advantage. A firm's strategic architecture influences its use of resources.
- Applying knowledge in action and learning are the foundations of a firm's competencies and capabilities.
- Firms function as open systems of resource flows motivated by managers' perceptions of the strategic gaps a firm must close to achieve an acceptable level of goal attainment. Firms have distinctive strategic goals that lead to unique patterns of resource flows and competence building and leveraging activities.
- Competence leveraging drives short-term competitive dynamics, while competence building drives long-term competitive dynamics.
- The complexity and uncertainty inherent in managing resource flows in a dynamic environment make the 'contest between managerial cognitions ' in devising strategic logics a primary feature of competence-based competition.
- Firms rely on the use of both firm-specific and firm-addressable resources, and competition occurs in markets for key resources as well as in markets for products.
- Competence-based competition includes forms of cooperation (as well as competition) with providers of key resources.
- Firms' differing abilities in coordinating resources and resource flows and in managing their systemic interdependencies greatly influence competitive outcomes in dynamic environments.
- Creating a systemic organizational capacity for strategic flexibility may be the dominant logic for competence-based strategic management in dynamic environments.

Resource-based view

- Firm growth is motivated by the availability of the firm's resources.
- Firm growth is limited by management's recognition of productive opportunities suited to the firm's available resources. The ability to combine existing and new resources; and a willingness to accept the risk of using new resource combinations will allow a firm to meet new market demands.
- Resource position barriers can be created when experience in using resources lowers costs for incumbents and imposes higher costs on imitators.
- Diversification is an attempt to extend a firm's resource position barrier into new markets by combining its current resources with new resources.
- Mergers and acquisitions are attempts to acquire groups of attractive resources.
- Firms cannot create a sustained competitive advantage in markets with homogeneous and perfectly mobile resources.

Table 2.2. Continued.

- Creating a sustained competitive advantage depends on control of a firm resource endowment that includes resources that are heterogeneous, imperfectly mobile, valuable, rare, imperfectly imitable and non-substitutable.
- The rent-earning potential of resources results from the properties of resources that create asset mass efficiencies, asset mass interconnectedness and time compression diseconomies in firms' efforts to accumulate and to create assets.

Dynamic capabilities framework

- Changes in economic activities result from the learning and embedding of new skills in new organizational routines.
- Skill development in organizations follows natural trajectories determined by the organization's existing skill base and routines.
- Competitive advantage arises from a firm's current distinctive ways of coordinating and combining its difficult-to-trade and complementary assets.
- At any point in time, certain assets will be important determinants of a firm's ability to earn rents in a given market, i.e. they will be strategic industry factors, but these assets will be imperfectly predictable and subject to market failure.
- Managers' cognitive and social processes will determine the assets a firm acquires and thus its potential for generating organizational rents.

propositions in the competence perspective in general, and the resource-based view and the dynamic capabilities framework, in particular.

2.6.1 Resource-based view

The resource-based view of the firm (RBV) is an influential theoretical framework for understanding how competitive advantage within firms is achieved and how that advantage can be sustained over time (Schumpeter, 1934; Penrose, 1959; Wernerfelt, 1984; Prahalad and Hamel, 1990; Barney, 1991; Nelson, 1991; Peteraf, 1993; Teece *et al.*, 1997; Scholten, 2006). This perspective focuses on a firm's internal resources and how these are acquired from factor markets, e.g. the labor and financial markets. In contrast to the industrial perspective that views resources as immediately accessible, the RBV stresses the inherent immobility or stickiness of valuable factors of production and the time and cost required to accumulate those resources (Peteraf, 1993). This causes firms to be idiosyncratic because throughout their history they accumulate different physical assets and, often more importantly, acquire different intangible organizational assets of tacit learning and dynamic routines (Dosi, 1988). Competitive imitation of these assets is only possible through the same time-consuming process of irreversible investment or learning that the firm itself underwent (Dierickx and

Cool, 1989). The irreversible investments made in these assets function as commitments that deter the duplication of valuable product-market positions and secure the distinctive value of the firm (Ghemawat, 1991). Such assets also produce 'path dependency' (Dosi *et al.*, 1988). So a firm's history, strategy and organization combine to yield the unique bundle of resources it possesses. Had it made different decisions in the past, its path of asset accumulation and hence the firm today would be different. Moreover, the future strategy of the firm is determined by its history, and its strategy is constrained by, and dependent on, the current level of resources (Collis, 1991). As a consequence, in RBV thought, Chandler's (1962) famous adage 'Structure follows strategy' is merely reversed to 'Strategy follows structure'.

In short, in the resource-based view, competitive advantage rests within the firm's idiosyncratic and difficult-to-imitate resources. A resource refers to an asset or input to production (tangible or intangible) that an organization owns, controls or has access to on a semi-permanent basis (Helfat and Peteraf 2003). It follows that a firm's resources include all those attributes that enable it to conceive of and implement strategies. They can be divided into four types: financial resources, physical resources, human resources and organizational resources (trust, teamwork, friendship and reputation). In RBV the firm's resources must be unique in four ways: Add Value. The resources must enable the firm to exploit external opportunities or neutralize external threats. Be rare. Ideally, no competing firms possess the resources. Inimitable. Competitors should not be able to imitate the resource either by duplicating it or by developing a substitute resource. For example, fast food discount coupons are a very poor competitive resource because competitors can quickly and easily print their own (duplication) or simply offer a temporary lower price (substitution). However, another set of barriers impedes imitation in advanced industrial countries. This is the system of intellectual property rights, such as patents, trade secrets, and trademarks. Intellectual property protection is not uniform across products, processes, and technologies, and is best thought of as an island in a sea of open competition. It presents an imitation barrier in certain contexts, although one should not overestimate its importance (Omta and Folstar, 2005). Ability to exploit. The firm should have the systems, policies, procedures, and processes in place to take full competitive advantage of the resources.

RBV builds on two basic assumptions about the firm's resources: (1) that they can vary significantly across firms (assumption of resource heterogeneity) and (2) that their differences can be stable (assumption of resource immobility). Furthermore, RBV considers firms to be rent-seekers rather than profit maximizers (Rumelt, 1987). Rent can be defined as the excess return to a resource over its opportunity costs. In other words, the payment received above and beyond that amount necessary to retain or call the resource into use. Rent-seeking behavior therefore emphasizes the important role of entrepreneurship and innovation. Firms continuously seek new opportunities to generate rents rather than contenting themselves with the normal avenues for profit. If control over scarce resources is the source of economic profits, then it follows that such issues as skill acquisition, the management of knowledge and know-how (Shuen, 1994) and learning become fundamental strategic issues. The more tacit the firm's knowledge, the harder it is for its competitors to replicate it. When the tacit component is high, imitation may well be impossible.

2.6.2 Dynamic capabilities framework

Teece *et al.* (1997) extended RBV to dynamic markets. The rationale is that RBV has not adequately explained how and why certain firms have a competitive advantage in situations of rapid and unpredictable change. In these markets, where the competitive landscape is shifting, the dynamic capabilities by which firm managers 'integrate, build, and reconfigure internal and external competencies to address rapidly changing environments' become the source of sustained competitive advantage (Teece *et al.*, 1997). In the dynamic capabilities framework (DCF), competitive advantage derives from a combination of competencies (skills and knowledge) with the managerial and technical systems (capabilities) that exploit those reservoirs in delivering value to customers (Leonard-Barton, 1995).

According to Eisenhardt and Martin (2000), dynamic capabilities consist of specific strategic and organizational processes, like product development, and alliancing that create value for firms within dynamic markets by manipulating resources into new value-creating strategies. These capabilities exhibit commonalities across effective firms in what can be termed 'best practices'. These include the local abilities or 'competencies' that are fundamental to a firm's competitive advantage such as skills in molecular biology for biotech firms or in advertising for consumer products firms. Dynamic capabilities are the antecedent organizational and strategic routines by which managers alter their resource base by acquiring resources and integrating and recombining them to generate new value-creation (Grant, 1996; Pisano, 1994). As such, they are the drivers behind the creation, evolution, and recombination of resources into new sources of competitive advantage (Henderson and Cockburn, 1994; Teece *et al.*, 1997). Dynamic capabilities thus are the organizational and strategic routines by which firms achieve new resource configurations as markets emerge, collide, split, evolve, and die. Hadjimanolis (2000) refers to the dynamic capabilities as 'the features of the firm and managerial skills forming organizational routines, which lead to competitive advantage'.

Teece *et al.* (1997) said that 'Winners in the global marketplace have been firms that can demonstrate timely responsiveness and rapid and flexible product innovation, coupled with the management capability to effectively coordinate and redeploy internal and external competencies'. They refer to the ability to achieve new forms of competitive advantage as 'dynamic capabilities' to emphasize two key aspects. The term 'dynamic' indicates the capacity to renew competencies and capabilities so as to achieve congruence with the changing business environment. Certain innovative responses are required when time/timing-to-market are critical. The term 'capabilities' emphasizes the key role of strategic management in appropriately adapting, integrating and reconfiguring internal and external organizational skills, resources and functional competencies to match the requirements of a changing environment.

Dynamic capabilities, as defined by Teece *et al.* (1997), build, integrate, or reconfigure a firm's competencies and capabilities. Zollo and Winter (2002) note that dynamic capabilities consist of routines. For example, a dynamic capability such as post-acquisition integration is composed of a set of routines that integrates the resources and capabilities of the merged firms (Capron and Mitchell, 1998). Another example is the product development routines by which managers combine their varied skills and functional backgrounds to create revenue-reducing products and services (e.g., Clark and Fujimoto, 1991; Dougherty, 1992; Helfat and Raubitschek, 2000). Similarly, strategic decision-making is a dynamic capability in which managers pool their various business, functional and personal expertise to make choices that shape the major strategic moves of a firm (e.g. Fredrikson, 1984; Judge and Miller, 1991; Eisenhardt and Martin, 2000).

2.7 The concept of strategic alignment

The concept of strategic alignment has played a key role in the development of strategic management thought (e.g. Zajac *et al.*, 2000). One of the most widely shared assumptions in the strategy literature is that the appropriateness of a firm's strategy can be defined in terms of the alignment - also referred to as fit, match, coalignment or congruence - of its strategy with both its external and its internal contingencies (Burns and Stalker, 1961; Lawrence and Lorsch, 1967; Ginsburg and Venkatraman, 1985; Miles and Snow, 1994; Verdú Jover *et al.*, 2005). The strategic alignment paradigm asserts the necessity of maintaining a close and consistent linkage between the firm's strategy and the context within which it is implemented (e.g. Venkatraman, 1989; Katsikeas *et al.*, 2006). The core proposition is that matching strategy with the environment leads to superior performance (e.g. Venkatraman and Prescott, 1990; Lemak and Arunthanes, 1997; Lukas *et al.*, 2001).

External fit demands that firms match their strategy with the opportunities and threats provided by the business environment, whereas internal fit requires the chosen strategy to be in compliance with the firm's internal structures and processes (Lawrence and Lorsch, 1967; Thompson, 1967). The process of strategic alignment is inherently dynamic, because strategic choices made by the firm will inevitably evoke counteractions (e.g. imitation, own innovations) by its major competitors, which will in turn necessitate a subsequent response. Thus, strategic alignment is not an event but a process of continuous adaptation and change (Henderson and Venkatraman, 1993). For the present study we define strategic alignment as follows.

Strategic alignment is finding the balance between the relevant contingencies in the business environment (external fit) and the firm's internal resources, competencies and capabilities (internal fit).

It is clear that a competitive advantage can be reached by creating superior strategic alignment. We therefore suggest that the process of strategic alignment is a capability in itself and should be contrasted with a firm's underlying technological competencies and managerial capabilities.

For a technology-based firm, the process of aligning its innovation strategy to its external environment and its internal resources, competencies and capabilities is so essential that it can be considered a core dynamic capability.

Miles and Snow (1994) proposed a strategy typology that interrelates organizational strategy, structure and process variables within a theoretical framework of alignment. They viewed the 'adaptive cycle' characterizing this process as involving three imperative strategic 'problem and solution' sets:

- an entrepreneurial problem set centering on the definition of an organization's product-market domain;
- an engineering problem set focusing on the choice of technologies and processes to be used for production and distribution; and
- an administrative problem set involving the selection, rationalization and development of organizational structure and policy processes.

The concept of alignment (or fit) is rooted in the population ecology model (Aldrich, 1979) and in the contingency theory tradition (Van de Ven *et al.*, 1989) and has played a pivotal role in the initial work in the field of strategic management, for example in the work of Schendel and Patton (1978). Venkatraman and Camillus (1984) distinguished two dimensions by which studies dealing with strategic alignment can be classified: (1) the strategic perspective (outside-in, inside-out or an integration perspective); and (2) whether the focus is on the content (what should be aligned) or on the processes (what actions should be taken to achieve alignment). Combining these two dimensions leads to a six-cell matrix in which each cell represents a different perspective on alignment in strategic management, explores different themes, and roots them in different theoretical streams (see Table 2.3). Research that focuses on the content of strategy has attempted to specify the strategic actions to be taken to match different environmental conditions (e.g. Chandler 1962; Ansoff, 1965; Andrews 1971; Porter, 1980). The group of content-oriented schools of thought views strategy as one of the system elements that has to be fitted to the other elements, such as the environment or the company's internal structures, or both. Research focusing on the process of alignment views strategic alignment as a continuous pattern of interactions aimed at achieving a dynamic match between the organization and its environment (e.g. Chakravarthy, 1982; Evered, 1983). The other dimension, with which Venkatraman and Camillus classify research on alignment, addresses the strategic perspective used. Some researchers focus primarily on strategy formulation aimed at creating alignment between external (market structure related) variables and strategic (firm conduct) variables, with no direct reference to the firm's internal resources, competencies and capabilities, i.e. the work of Rumelt (1982) who related diversification strategy and market structural variables to performance differences. Others are mainly concerned with strategy implementation issues and focus on how strategy can be aligned using internal structure (Galbraith and Nathanson, 1979), management systems (King, 1978), and organizational culture variables (Schwartz and Davis, 1981). The integration school argues that, in a

Table 2.3. Key issues and theoretical streams concerning strategic alignment. Adapted from Venkatraman and Camillus (1984).

Domain	Content approach	Process approach[1]
External	Strategy formulation school	Interorganizational (strategy) network school
	Aligning strategy with environmental conditions. • Industrial organization theory • Business policy/strategic management	Strategy analysis at the 'collective' level, emphasizing interdependence of strategies of various organizations vying for resource allocation. • Interorganizational networks • Resource-dependency themes • Constituency analysis
Internal	Strategy implementation school	Strategic choice school
	Tailoring administrative and organizational mechanisms in line with strategy. • Business policy • Normative strategy literature	Managerial discretion moderating the 'deterministic' view regarding decisions on organizational mechanisms. • Organization theory • Business policy-organization theory interface
Integrated	Integrated formulation-implementation school	Integrated network-process school[2]
	Strategic management involving formulation and implementation and covering both organizational and environmental decisions. • Business policy/strategic management • Markets and hierarchies program	Broadly configuring organization and environment, emphasizing interdependence but *not* causation. • Organizational theory • Business policy/strategic management • Population-ecology-based concepts

[1]Referred to by Venkatraman and Camillus as 'Pattern of interaction'.
[2]Referred to by Venkatraman and Camillus as 'Overarching Gestalt School'.

multi-industry context, both environmental and organizational variables are to be considered and strives for a synthesis of the two former perspectives.

Considering that the present study seeks to integrate the outside-in and the inside-out approaches to strategy, the present study should be placed in the integrated domain of this framework. The first empirical study (the cross-industry study) can be classified as being in the integrated formulation-implementation school, in line with the observations by Miles and Snow (1980) who argue that an organization's internal structures and management practices continually have to achieve an optimal level of fit with its environment. In the integrated network-process school, strategy is viewed as a process of aligning the elements that are partly internal and partly external to the organization. The concept of alignment in the latter school is along the lines of the dynamic, process-oriented interpretation of fit as suggested by Drazin and Van de Ven (1985). The theoretical support for this cell is derived from the open system perspective of organization theory (Katz and Kahn, 1966; Thompson, 1967) and the ecological view of organization and environmental transactions (Thorelli, 1977). In this perspective, strategy is perceived as the dynamic combination of environmental forces and internal resources, competencies and capabilities that continually affect each other.

2.8 Concluding remarks

The previous sections described the main theoretical perspectives that can be used to analyze a firm's strategy and the alignment of its strategy to its external and internal environment: the industrial organization and the competence perspective. We also discussed the main schools of thought in the analysis of strategic alignment, concluding, that the industrial organization theory proposes the markets and industry in which a company operates as being the main factors to analyze when investigating the strategic alignment of innovation to business, whereas the competence perspective proposes the firm's own resources, competencies and capabilities as the key factors to study. In our empirical studies we seek to integrate these two perspectives on strategy. The first empirical study, the cross-industry study that focuses on the content of alignment, can be classified in the integrated formulation-implementation school, while the second empirical study, the longitudinal study that focuses on the process of achieving alignment, can be classified as being in the integrated network-process school. Finally, we conclude that, for technology-based firms, the process of aligning the innovation strategy to a firm's external environment and its internal resources, competencies and capabilities can be considered as a core dynamic capability.

3. Study domain: innovation

In this chapter, we explore how the two theoretical perspectives on strategy and the two ways of analyzing the phenomenon of strategic alignment, which were identified in the previous chapter, can be applied to innovation, the domain of the present study. We start by introducing the phenomenon of innovation in Section 3.1. In Section 3.2, different typologies of innovation are discussed based on the object of innovation and the level of newness and/or disruptiveness. In Section 3.3 we elaborate on the R&D process, which is so important for a technology-based firm's long-term survival in the market. We consider the stage gate R&D funnel concept and first to fifth generation R&D. Section 3.4 discusses the barriers to innovation and Section 3.5 then considers the drivers of innovation. We focus on the innovation culture and strategy of the technology-based firm and the organizational setting in which innovations are produced, either in-house (closed innovation) or with third parties, for example suppliers, buyers, competitors or knowledge institutions (open innovation). In Section 3.6 best practices as derived from the literature are discussed in some detail. Section 3.7 looks at the way an innovation strategy is formulated and implemented in technology-based firms, and elaborates on how innovation strategy is viewed by the two perspectives on strategy, as discussed in Chapter 2. Section 3.8 reviews how the two approaches to strategic alignment (identified in Chapter 2) can be applied to innovation strategy. The chapter ends with some concluding remarks in Section 3.9.

3.1 The phenomenon of innovation

Over the years, the subject of innovation has been studied from two broad perspectives. The first, an economics-oriented tradition, examines differences in the pattern of innovation across countries and industrial sectors, the evolution of technologies and inter-sectoral differences in innovation (e.g. Rosenberg, 1982; Dosi *et al.*, 1988; Nelson, 1993; Niosi, 1995). The second, management-oriented tradition focuses on the micro- and meso-level and how new products are developed. These studies differ with respect to the sector studied, the level of aggregation (individuals, projects, firms or inter-firm innovation), the size or type of company (high-tech start-ups, large conglomerates.), the scope (incremental or radical, disruptive or sustaining innovations) or type of innovations studied (product, process or organizational innovations) and the geographical setting.

The popularity and wide applicability of the word 'innovation' has resulted in a proliferation of its meanings. In this book we start from the broad definition of innovation as provided by Schumpeter (1934):

The introduction of a new good -that is one with which consumers are not yet familiar- or of a new quality of a good. 2) The introduction of a new method of production, which need by no means be founded upon a discovery scientifically new, and can also exist in a new way of handling a commodity commercially. 3) The opening of a new market that is a market into which the particular branch of manufacture of the country in question has not previously entered, whether or not this market has existed before. 4) The conquest of a new source of supply of raw materials or half-manufactured goods, again irrespective of whether this source already exists or whether it has first to be created. 5) The carrying out of the new organization of any industry, like the creation of a monopoly position (for example through trustification) or the breaking up of a monopoly position.

This definition implies that innovation means more than just the creation of new products, processes and services and may also include innovation of business models, management techniques and strategies and organizational structures (Hamel and Prahalad, 1994). Innovation typically involves creativity but is not identical to it: innovation involves acting on creative ideas to make some specific and tangible difference in the domain in which the innovation occurs. For example, as Amabile (1996) stated: *Creativity by individuals and teams is a starting point for innovation; the first is necessary but not sufficient condition for the second.* It is important to note that innovation is not the same as invention. In general, an invention refers to the result of research activities (e.g. a patent), while an innovation is a commercial product, process or service. Martin (1985) describes it as follows: *An invention may be viewed as a new idea or concept, but this invention only becomes an innovation when it is transformed into a socially usable product.*

The knowledge needed to create innovations can be either stored in media (explicit knowledge, for example, specifications, procedures, reports and patents), or in people's heads (tacit knowledge, for example, trade secrets based on know how, Nonaka, 1994). Tacit knowledge finds its basis in expertise, skills and creativity. Craftsmanship and experience usually have a large tacit component and important aspects of innovation may not be expressed or codified in manuals, routines and procedures, or other explicit articulations (Burgelman and Rosenbloom, 1997). As Davenport and Grover (2001) argue, knowledge is context specific. Even when you know how a firm does something, it is very hard to replicate it. For instance, many automotive companies around the world have had great difficulty implementing the Toyota Production System from the 1980s, even though its principles have been widely published (e.g. Clark and Fujimoto, 1991). Innovations are difficult to implement by other firms because they do not come with sufficient context to allow successful application. For a recent extensive overview of the innovation articles published in the leading journal of Management Science the reader is referred to Shane and Ulrich (2004).

3.2 Innovation typologies

The various meanings that the term 'innovation' has acquired over the years can be clustered into three main concepts (Zaltman *et al.*, 1973). *The new item itself* refers to the object of innovation, i.e. the new or improved product, service, process or management technique. *The process of diffusion of the new item* is the process of user acceptance and implementation that was extensively studied by Rogers' and described in his classic book, 'The Diffusion of

Innovations' (1995). And *the process of developing the new item* is the concept that refers to the innovation process itself. It is also referred to as the R&D process in technology-based firms, starting with the innovative idea based on technology push research or market demand and developing it into widespread utilization. Looking back on research done so far, we could state that the dominant focus in innovation research has gradually evolved from the new item itself and the process of adopting the new item towards the process of developing it. In the 1960s and 1970s, studies on the 'innovation' phenomenon concentrated primarily on the diffusion of innovation (Rogers, 1995) or on the technical aspects of innovation (Roussel *et al.*, 1991). It was not until the mid 1980s that emphasis was put on the organizational systems in which innovation processes were taking place. The characteristics of the innovative firm were studied (e.g. Moss Kanter *et al.*, 1997) and management research identified managerial and organizational factors that enhanced or inhibited the success of innovations. Different types of innovation can be identified based on the object of innovation, including *product and service* innovations on the one hand, and *organizational* innovations on the other.

Product and service innovations involve the introduction of products or services to the market that are new or substantially improved. These may include improvements in functional characteristics, technical abilities and ease of use. There are several types of new products and services. Some are minor modifications of existing products, while others are completely new to the firm, the market or even the world. Product and service innovations are, therefore, often positioned either according to their level of newness in *incremental* (or evolutionary) innovation versus *radical* (breakthrough) innovation. While incremental innovations involve the adaptation, refinement and enhancement of existing products and services with a high chance of success and low uncertainty about outcome, radical innovations involve leaps in the advancement of a technology or processes leading to entirely new products, processes and services. Incremental innovations include *me-too products, line extensions, and repositioned products*. A me-too product is basically the same as a product that is already on the market, but produced by another company. A line extension is a variant of an existing product, produced by the same company that implies only small changes in manufacturing, marketing, storage and handling. A repositioned product is a product that is promoted differently then the existing product, e.g. to capitalize it in a certain niche market.

Christensen (1997) proposes a typology according to the level of market disruptiveness by positioning *sustaining* versus *disruptive* innovations. Sustaining innovations refer to the successive, sometimes important, technological improvements building on existing technologies that allow firms to continue to approach markets in the same way, such as the development of a faster or more fuel-efficient car. Disruptive innovations, by contrast, typically build on radical new technologies and, despite the fact that these technologies may often initially perform worse than existing mature technologies, they can eventually surpass them by either filling a role in a new market that the older product could not fill (e.g. laptop computers in the 1990s) or by successively moving up-market through performance improvements until finally taking over the whole market (e.g. the rapid replacement of film photography by digital photography).

Organizational innovations involve the creation or alteration of business structures, practices and models, and may therefore include process, supply chain and business model innovation (e.g. Carr, 1999).

- Process innovations involve the implementation of new or significantly improved production and manufacturing methods.
- Supply chain innovations are innovations that occur in the sourcing of inputs from suppliers and the delivery of output to customers.
- Business model innovations involve changing the way business is done in terms of how a company plans to serve its customers (the customer-value proposition) and how it plans to organize its activities.

3.3 The R&D process

We now elaborate on the process of developing new items, which is the focus of the empirical studies in this research project. Especially in technology-based firms, the R&D process constitutes a very important condition for a firm's survival in the market. In the Frascati Manual, Research and Development (R&D) is defined as follows (OECD 1994):

Creative work undertaken on a systematic basis in order to increase the stock of knowledge, and the use of this stock of knowledge to devise - ... - new materials, products, or devices - ... - new processes, systems or services, or - ... - improving substantially those already produced or installed.

The OECD (1994) distinguishes between three types of R&D activities: basic research, applied research and experimental development.

Basic (fundamental) research is defined as original investigation undertaken in order to gain new scientific and/or technical knowledge and understanding (Freeman, 1982).

Basic research is often pursued in corporate R&D as well as in university research centers. This form of R&D feeds the value chain for new product development by making scientific discoveries and earns a return on investment by claiming ownership to intellectual property through patents and proprietary knowledge. Basic research is often related to curiosity and the urge to discover and elucidate new and unconcealed phenomena. Researchers are led by their own ideas and scientific interests or those of their direct supervisor(s). Basic research is also connected with serendipidity. This means that important discoveries are often made as accidental side-products of research directed towards other subjects. For instance, Aspartame, a sweetener used in many food products, was a chance discovery. Because basic research can be highly uncertain and risky, from a business perspective it is hard to justify investment unless there is some clear idea of the potential market value of new knowledge discoveries. We refer to this kind of research as 'applied research'. Freeman and Soete(1997) defines it as follows:

Applied research is undertaken to gain new scientific and/or technical knowledge, but it is directed primarily towards a specific practical aim or objective.

Most of R&D budgets are spent on technology development and commercialization. Development activities are thus increasingly conducted in a parallel and yet integrated way. Formerly, research laboratories working on different parts of the R&D process were sequentially dependent in a chain of R&D activities. They used the results of an upstream department, transformed them, and passed them through to a downstream department. Communication between the different departments was limited. When intensified competition forced the companies to accelerate their R&D process, the linear sequence was gradually replaced by parallel development. Downstream activities started before having received finalized information from upstream R&D activities. However, because communication between upstream and downstream departments has not intensified, integration problems have arisen. Allen (1977) found a high level of association between the flow of information between scientists in different phases of the R&D process and the performance of an industrial laboratory. In recent years, companies have markedly intensified communication across the whole R&D process and in marketing and production (lateral and cross-functional communication), leading to concurrent development. In accordance with Allen's findings, upstream and downstream activities are both benefiting from this improved communication and integration (Clark and Fujimoto, 1991). A functional hierarchy does not support lateral and cross-functional communication. To achieve that, project goals have to take precedence over functional goals. Many companies have installed lateral and cross-functional project teams that draw on members from throughout the organization (Donnellon, 1993; Henke *et al.*, 1993).

3.3.1 The R&D funnel

The R&D funnel is the familiar image of the R&D process. A large number of innovative ideas enter the 'mouth' of the funnel. These ideas flow towards the 'neck' of the funnel where many will be eliminated. The neck can be loosened or tightened depending on the innovation strategy and the availability of development teams within a firm or in cooperation with other firms and/or knowledge institutions. The selected ideas gradually proceed through the different development phases until they are launched on the market. Phased development processes, today frequently called stage gate processes, break the R&D funnel up into time-sequenced stages separated by go/no go/adapt-management-decision screens between the phases (Cooper *et al.*, 2001, see Figure 3.1). Several stages can be distinguished in the R&D funnel.

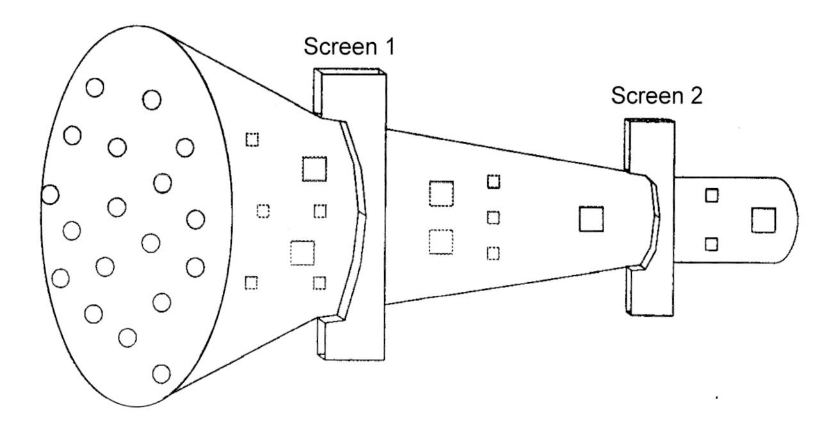

Figure 3.1. The R&D funnel.

In the *idea generation phase,* the environment is scanned for new ideas. This phase is often referred to as the 'fuzzy front end', to indicate the messy 'getting started' period in the R&D process. It parallels the invention phase in Schumpeter's well-known invention - innovation - commercialization trilogy (Schumpeter, 1934). Innovative ideas can be obtained from marketing, R&D, competitors or customers, and idea-generating techniques such as brainstorming are typically used in this phase. Although this phase may not be the most expensive part of the R&D process, it is here that the major commitments are made involving time, money, and the product's nature (Smith and Reinertsen, 1998). In the *idea screening phase,* ideas are screened for technical feasibility, cost and customer value. In this phase the question as to how to protect the property rights has to be answered. Several legal concepts may apply to any given innovation, product, process or creative work. These include patents, trademarks, trade names, copyrights and trade secrets. In the *concept development and testing phase*, the marketing and engineering focus is detailed by describing the target market, the product benefits and any manufacturing challenges. Virtual development and rapid prototyping techniques are increasingly being used to speed up development. In the *business analysis and beta (market) testing phase,* the expected sales volume, selling price and break-even point are established and a physical prototype or mock-up is produced and tested to determine customer acceptance. In the *commercialization phase* the product is launched, promotion material is produced, a new supply chain is built, if necessary, and the distribution pipeline is filled.

3.3.2 First through fifth generation R&D

Roussel *et al.* (1991) described three generations of R&D management practice from the 1950s until the early 1990s. In First Generation R&D in the 1950s, R&D was basically technology driven, the R&D phases followed each other sequentially, and there was less attention to the market. R&D was perceived as an overhead and managed as a traditional, hierarchical, functionally driven organization. In Second Generation R&D in the 1960s and 1970s, R&D departments began to link up with other business functions. Increased interdependence fostered cooperation and led to an increased focus on the market. R&D processes were increasingly led by strategic forecasting. By the 1980s Third Generation R&D had arrived and R&D management sought to reach across the entire enterprise, creating formal linkages with business units. R&D management became more systematic, with general and R&D managers jointly exploring and determining technology portfolio decisions. Miller and Morris (1999) introduced Fourth Generation R&D that included a process of integration of the different phases in the R&D process. Shortening time-to-market was essential because risk had to be balanced with business opportunity, which decreased over time. In recognizing the need for cross-functional and cross-disciplinary insight, emerging 'communities of practice' became integral to understanding future business opportunities. In both Third and Fourth Generation R&D, customer satisfaction was the focus. In 1993 Rothwell already anticipated the arrival of Fifth Generation R&D in the next millennium. The ideas behind it were translated by Chesbrough in his famous concept of open innovation in 2003 that is described in the next sub-section. To cope effectively, Rothwell envisioned R&D management systems having to be knowledge-based and directed to networking with suppliers, distributors, customers and other stakeholders.

3.4 Barriers to innovation

Few innovative ideas prove profitable because the research, development, and marketing costs of converting a promising idea into a profitable product are extremely high. A study by Booz-Allen and Hamilton (1982) of more that 700 US manufacturers showed that less than 2% of the innovative projects initially considered by 51 companies eventually reached the market place. To be more specific, out of every 58 new product ideas, only 12 passed an initial screening test that found them compatible with the firm's mission and long-term objectives, only 7 remained after an evaluation of their potential, and only 3 survived development. Of these 3 survivors, 2 appeared to have profit potential after test marketing and only 1 was commercially successful. Most of the commercialization failures occurred because the idea or its timing was wrong. The American Product Development and Management Association (PDMA) sponsored an effort to describe developments since Booz-Allen and Hamilton's 1982 study. This new study found that the mortality rate of products proceeding through development had increased only slightly since 1982: one successful product resulted from 11 product ideas or concepts that passed initial screening, versus 12 in the earlier study (Griffin and Page, 1993; Hollander, 2002).

Innovations that fail are often potentially good ideas but have been rejected or shelved due to budgetary constraints, a lack of skills or poor fit with current goals. Early screening avoids unsuitable ideas devouring scarce resources that are needed to progress more beneficial ones. In fact, failure is an inevitable part of the innovation process and most successful innovative firms expect a certain level of failure. While learning from failure is important, failure rates that are too high are wasteful and a threat to a firm's future (Cobbenhagen, 1999; Huizenga 2000). Much attention has been consequently directed to the barriers to innovation. According to the AMA 2005 Innovation Survey, the top three barriers are insufficient resources, the lack of a formal strategy for innovation, and a lack of clear goals and priorities. Also important are organizational structures that are not geared to enhancing innovation. In a similar vein, a Conference Board study of 100 firms (Troy, 2004), primarily from the USA and Europe, looked at barriers to innovation success and found that among the most commonly cited ones were a lack of organizational alignment (52%), insufficient resources to pursue new ideas (51%), no formal innovation strategy (49%), and a lack of goals and measures (44%).

3.5 Drivers of innovation

In the next sub-sections the most important drivers of innovation are discussed. First the questions of how to create a culture in which innovation can flourish and how to craft a strategy directed to innovation are addressed, and then the management systems directed to open innovation are highlighted.

3.5.1 Innovation culture

An organization's potential to unleash the creativity of its members is to a great extent determined by its innovation culture (Senge, 1990). Innovation requires experimentation and thus a tolerance for failure and a redundancy or 'slack' in resources (see, for example, Rosner, 1968; Subramanian, 1996; Gopalakrishnan, 2000). While certainly capable of conducting thorough and highly sophisticated research, large technology-based firms are often somewhat bureaucratic and may lack the flexibility and entrepreneurial drive that are so characteristic of small entrepreneurial companies. For this reason, many technology-based firms work closely with small companies (*knowledge acquisition*) and/or try to create an entrepreneurial climate within their companies (often referred to as intrapreneurship). To achieve this, the exchange of staff is promoted, and employees are encouraged to come up with new ideas through idea boxes, incentive systems, brainstorming workshops or providing employees with innovation or scouting time (e.g. Hüsig and Kohn, 2003).

Kuczmarski and Associates (1994) published a study based on 77 respondents in a cross-section of industries in which it appeared that successful companies showed more tangible and visible signs of *top management commitment* to innovation, especially in terms of providing adequate funding and resources. They also focused more effort on new-to-the-world and new-to-the-company products, devoted a larger percentage of the R&D process to concept screening and testing and rated themselves as being effective in terminating projects during development. Other studies have also shown that if top management is visibly and tangibly committed to innovation, R&D is clearly more successful (Arthur D. Little, 1991; Bart, 1991; Mercer Management Consulting, 1994).

3.5.2 Innovation strategy

Miles and Snow (1978) define four types of generic strategies that a firm can pursue: prospector, analyzer, defender or reactor. Their theory holds that, in order to be superior, there must be a clear match between the firm's mission and values, its corporate and functional strategies. Companies embracing a prospector strategy want to be at the forefront of innovation, and they seek to reap the initially high profits associated with customer acceptance of a new or greatly improved product. The underlying rationale of the prospector strategy is to create a new product generation life cycle and thereby make similar existing products obsolete. Prospectors react immediately to market opportunities and are often among the first to introduce innovations to the market. Yet, it is not necessarily always the prospector that will ultimately become the market leader. Particularly with completely new products, the advantages of being the 'first mover' often turn out to be illusory. After all, the prospector has to deal with higher development and marketing costs, while other firms can just copy the product and optimize it based on market experiences (me-too-but-better). For completely new products, the firms that are second or third to enter the market in a number of cases become the market leader (Hart *et al.*, 1998). Analyzer firms analyze and imitate the successes of their competitors. Analyzers operate in two types of product-market domains, one relatively stable, the other rapidly changing. In the stable areas, these organizations operate routinely and efficiently by using formalized structures and processes. In the more turbulent areas, top managers watch their competitors closely for new ideas and then rapidly adopt those which appear to be the most promising. Although seldom 'first-in', they are fast followers. It is not unusual to see analyzer firms develop the necessary technology and then wait with further developments until a competitor introduces the new product on to the market. A defender firm attempts to locate and maintain a secure market presence in a relatively stable product or service area. It tends to offer a more limited range of products or services than its competitors. A defender firm is usually not at the forefront of developments in the industry. Finally, a reactor firm lacks a consistent strategy and product/market orientation. It seldom makes adjustments of any sort until it is forced to do so by market competition.

Exploration versus exploitation

The choice of an innovation strategy also has to do with finding the right balance between exploration and exploitation. The distinction between exploration and exploitation was first noted by Holland (1975) and was later further developed by March (1991). Exploitation is associated with the refinement and extension of existing technologies, which adds to the competencies and capabilities of firms without changing the nature of their activities. As a consequence, exploitation can be planned and controlled, which is important as competition will already have emerged and considerations of efficiency are crucial. In contrast to exploitation, exploration is concerned with the experimentation with new alternatives and can generally be characterized by breaking from existing rules, norms, routines and activities to pursue novel combinations. Hence exploration is not about the efficiency of current activities and cannot be planned for. It is an uncertain process that deals with the constant search for new opportunities. Returns from exploitation are positive, proximate and predictable. By contrast, returns from exploration are uncertain, more remote in time and organizationally more distant from the locus of action (Levinthal and March, 1993). Performing both tasks is important for firms that operate in technology-based industries. Given the high rate of change that generally characterizes such firms (Hagedoorn, 1993), exploitation enables them to recoup rapidly the (large) investments made in existing technology. At the same time, these fast-changing conditions quickly make existing technology obsolete and this then requires the timely creation and development of new technology. In short, exploitation is needed in the short run, while exploration is required for the long-term survival of a technology-based firm.

Technology push versus market pull

Another important dichotomy of a company's innovative strategy depends on whether it is technology push or market pull. The innovation strategy finds its expression in the way it strikes the delicate balance between letting the technological possibilities or the market drive innovation. Although there is much empirical evidence that underlines the importance of a market-oriented innovation strategy (see next sub-section), different authors (e.g. Johne and Snelson, 1988) argue that technology push and market pull innovation should be treated as equal elements in an integrated innovation approach. Very often technology-market roadmaps are used to visualize this aim to balance technology push and market pull innovation (e.g., Albright and Kappel, 2003; McCarthy, 2003).

Market orientation

Cooper (1999) argues that one of the most important best practices is delivering differentiated products with unique customer benefits and superior customer value. Surprisingly, product superiority is often absent as a project selection criterion, while rarely are concrete steps built into the innovation process that encourage the design of superior products. Unfortunately, he argues, there is a too great an emphasis on R&D cycle time reduction and a tendency to favor simple, incremental projects, which actually penalizes R&D projects that could lead to product superiority. To reach superior product performance, a strong market orientation is of the utmost importance. Innovation projects that feature high-quality marketing actions -preliminary and detailed market studies, customer tests, field trials and test markets - show clearly higher success rates and a higher market share than projects with poor marketing research. The recent emphasis on market orientation has resulted in the increased integration of customer values early on in the innovation process. For example, the 2004 Conference Board study of 100 firms, primarily from the US and Europe, found that customers were major factors in the companies' innovation goals for 2006. Over 7 in 10 respondents rated the following goals as highly important: improving customer satisfaction via new processes (79%), increasing loyalty among current customers (73%) and identifying new customer segments (72%, Troy, 2004).

Not surprisingly, a strong market launch underlies successful products. Teams that develop new products successfully devote more time and money to the market launch. As a consequence, the quality of execution of the market launch is significantly better. In some businesses, it would seem that the launch is a major concern after the product is fully developed (Cooper, 1999).

User experience may provide important feedback for market-oriented innovation. As Mercer Management Consulting (1994) indicates, one of the practices that contributes most to differentiating between low- and high-performing companies involves including potential customers directly in the different stages of R&D. In this regard, Von Hippel (1988) emphasizes the importance of selecting the 'right' consumers to serve as the 'lead users'. Using a random selection of customers can at best lead nowhere and at worst push the innovation process in the wrong direction.

3.5.3 Open innovation

According to Chesbrough (2003), innovative companies increasingly realize that the 'closed' model of innovation, in which the internal R&D department exclusively provides for new products and processes to foster the company's growth, does not work any more in the current highly dynamic business environment. From the resource-based perspective, it has been argued that external networks have the potential to deliver a wide range of ideas, resources and opportunities far beyond the ability of the organization on its own (e.g. Ahuja, 2000; Gulati *et al.*, 2000; Hadjimanolis, 2000; Quinn, 2000; Duysters and Lemmens, 2003). As Quinn (2000) points out, in order to compete in current markets, cooperation within a network of partners is becoming more and more essential. As proven empirically by Caloghirou *et al.* (2004), interacting with external partners enables a firm to access a variety of new knowledge - a phenomenon they have termed 'enhanced absorptive capacity' - which increases innovative performance. The ability to identify potential network partners and maintain existing relations with current partners are thus of crucial importance.

Under the paradigm of open innovation, R&D results that otherwise would have gone unutilized are transferred across the firm's boundary, for example by out licensing to another company or in a joint venture, or by spinning out and launching a new venture that uses the technology,. Similarly, a firm may license in technologies created by other firms that are useful to its own core business. Recently, many technology-based firms have formed alliances with start-up firms and have built up their own internal venturing groups of senior managers who scout for new ideas, products and processes to fill the R&D pipeline. Huston and Sakkab (2006) refer to this new paradigm of open innovation as 'Connect and Develop', instead of 'Research and Develop'.

There is only limited research to support the idea of the beneficial impact of knowledge sharing in R&D but what does exist is compelling. Omta (1995) related R&D performance in the pharmaceutical industry to the openness of the organization's information and knowledge culture. He found that the best-performing pharmaceutical firms were characterized by less management concern about the leaking of company information and greater openness to outside information, including higher attendance at scientific conferences. Omta's research suggests that the more an organization wants to share its information with the external environment, the more it gets in return. A company that spends most of its energy hoarding and protecting its own knowledge will be less open to new knowledge from the outside world. Innovative companies are generally those that do not rest on their intellectual laurels but instead are constantly on the lookout for new innovative ideas they can use to develop new products, processes and services.

3.6 Best practices in innovation

Innovation research so far has uncovered two types of best practice, one directed to effectiveness in 'doing the right projects', and the other directed to efficiency in 'doing the right projects right'. The best practices directed to effectiveness are related to strategic project choice. They include scoring models of the potential of the new product's market, the existing and emerging technologies, and the competitive situation, along with the ability to leverage the firm's core competencies. An overview of best practices as revealed by different studies is given below.

3.6.1 Clear product definition and portfolio planning

The failure to define a product's target market - the concept, benefits, features and specifications - before development begins is a major cause of new product failure and may result in a serious delay in time-to-market. However, in the current dynamic business environment, new products often succeed or fail for unforeseen reasons. It is impossible to specify in advance how markets and rivals will react to an innovation, but it is possible to manage a portfolio of projects by employing R&D portfolio planning techniques (Roussel *et al.*, 1991; Cooper *et al.*, 2001). The purpose of R&D portfolio planning is typically to reach the optimum point between risk and reward, and stability and growth. The definition of optimum, however, varies widely depending on the circumstances and, in particular, on the interdependency of risk, uncertainty, technological maturity, technological impact and the competitive situation. R&D portfolio management is basically a dynamic decision-making process, in which a list of R&D projects is constantly updated and revised. In this process, new projects are evaluated, selected and prioritized, existing projects are accelerated, terminated or de-prioritized, and resources are (re-)allocated. Multiple approaches to R&D portfolio management have been developed over the past few years. These include techniques, such as financial methods, business strategy methods, bubble diagrams, scoring models and check lists. From the existing approaches, a combination of strategic and financial methods is most commonly used (Cooper *et al.*, 2001). A recent development in portfolio management is the use of real option theory to maximize option value. Real option theory basically perceives R&D opportunities as call options in which the corporation has the right, but no obligation, to invest (Luehrman, 1998). By basing the decision process not only on the net present value of R&D projects, uncertain and changing information and dynamic opportunities, multiple goals and strategic considerations can be taken into account enabling the effective linkage of the R&D portfolio to strategic planning (Luehrman, 1998; Zbignew and Pasek, 2002).

3.6.2 A structured R&D process

The decision-making process can be supported by a number of tools and methods. Project milestones are an important steering tool once projects are underway. Management can define in advance at what stages of the project the team has to hand over its partial results in order to gain insight into and control over the project. Key performance indicators (KPIs) can support decision-making (e.g. Omta and Bras, 2000). They constitute a set of indicators that Cordero (1990) refers to with the term 'control stage measures'. These are intended to record the outputs realized and the resources used, in order to provide insight into the efficiency of the process. A comparison of these measures to the standards estimated provides management with insight into the degree of under- and overestimation. The indicators 'within budget', 'on-time' and 'according to specifications' are the three most commonly used key performance indicators for project assessment (see, for example, Shenhar and Tishler, 2002).

In 1994 Mercer Management Consulting gathered survey responses from 193 R&D managers in a variety of industries and linked R&D practices to innovative performance, which was defined as combined self-assessments of cycle time, innovativeness, success rate and revenue contribution of new products (Mercer Management Consulting, 1994). They found that high performers were differentiated from lower performers in their execution of a commonly agreed to, customer-centered and disciplined new product development process, their cultivation of a supportive organization and infrastructure for new product development, and in setting the innovation agenda and managing the portfolio of projects in aggregate. In too many companies, projects move far into development without serious scrutiny: once a project begins, there is very little chance that it will be stopped. The result can be that many marginal projects are approved. Indeed, having tough milestones, go/no go decision points or stage gates correlates strongly with profitability of innovation. In 1995 Pittiglio Rabin Todd and McGrath (PRTM) used responses from over 200 organizations from six industry groups to determine best practices in new product development (Pittiglio *et al.*, 1995). They defined the 'best-in-class' as the top 20% against a set of six new product development matrices: time-to-market, time-to-profitability, project goal attainment, new product revenue contribution, and wasted development project spending (McGrath and Romeri, 1994). The study identified several best practices. At the project level, best practices include using cross-functional teams and a structured development process with action-oriented stage reviews.

However, the stage gate approach has been criticized by several authors (e.g. Smith, 2006) who argue that it fosters a mindset in which the R&D process proceeds sequentially so it becomes difficult to even conceive of a highly overlapped, iterative rugby-type process. A result of unexpected technical and market challenges, the innovation process will almost certainly be iterative and concurrent rather than unidirectional and sequential (Janszen *et al.*, 1999).

3.6.3 Use of (international) cross-functional project teams

Over the last decade, the importance of providing greater autonomy and responsibility to self-steering, empowered project teams has been stressed (e.g. Wheelwright and Clark, 1992; Von Zedtwitz, 1999; Hauser, 2001; Calantone *et al.*, 2002). In 1991, the Consultancy firm Arthur D. Little surveyed the product innovation processes of 701 companies in nine manufacturing industries, focusing primarily on top management's concerns and improvement efforts. His study indicated that projects that are organized using cross-functional teams that are responsible for all aspects of the project from initial idea generation to final commercialization have more chance of being successful because the innovation process typically requires the expertise of different functions, e.g. R&D, marketing, manufacturing and procurement (Arthur D. Little, 1991). An international orientation means defining the market as an international one and designing products to meet international requirements, not just domestic ones. An international orientation also involves adopting a transnational innovation process, utilizing cross-functional teams with members from different countries and gathering market information from multiple international markets as input.

3.6.4 Use of integrated virtual development tools

In the R&D process, management tools are playing an increasingly important role. The primary tools for project management were initially introduced in the 1950s and 1960s to increase R&D effectiveness. The effective use of project management tools usually has a strong impact on meeting project schedules and budget objectives. The impact on project success in terms of meeting either functional or technical specifications was much less predictable (Raz *et al.*, 2002). However, McGrath and Romeri (1994) indicated that currently an integrated set of virtual development tools (such as QFD, rapid prototyping, and modeling and simulation) are essential for R&D success.

However, Griffin (1997) concluded that most of the above mentioned studies disregard the contextual differences and suggested that more study is needed to better define best practices within contexts. For this reason, the cross-industry survey (see Chapter 4) takes the often large contextual differences between technology fields and industries (such as electronics, aircraft and pharmaceuticals) as the starting point for the analysis.

3.7 Innovation strategy and the two perspectives on strategy

In Section 2.1 we identified four phases in the process of crafting a strategy: strategic analysis, strategy formulation, strategy implementation and execution, and strategy evaluation. We will now use this framework to discuss the crafting of an innovation strategy. In the *strategic analysis phase* (that includes *scanning* and *idea generation*), signals from the environment about potential opportunities and threats are detected. As Drucker (1985) argues:

Systematic innovation [...] consists in the purposeful and organized search for changes, and in the systematic analysis of the opportunities such changes might offer for economic and social innovation. The more firm management is alert to opportunities for innovation, the more likely new ideas are recognized and adopted within the organization.

As the former CEO of Hewlett-Packard, Patterson (1993) stated, the moment a business opportunity occurs is often followed by a long delay until the opportunity is perceived by the company. It is management's job to scan the environment so that the delay is kept to a minimum. The processes in the strategic analysis phase can be facilitated by employees fulfilling 'boundary spanning' and 'gate keeping' roles (Brown and Eisenhardt, 1995; Reid and De Brentani, 2004). In the *strategy formulation phase,* ideas and opportunities are linked to a firm's strategy, leading to selecting those that fit into its strategy and abandoning those that do not. Once an innovation strategy has been crafted and decided upon, it must be implemented, i.e. translated into action. This comprises a process of planning (implementation) and actions (execution), with decision-making taking place throughout the process and at gradually lower levels in the organization. At project level, choices have to be made by the R&D management, for example regarding resource allocation, which project proposals to approve, and which projects to continue, stop or alter. Decisions in the execution phase can also be taken by the team conducting the innovation project. In the *evaluation phase,* the information gathered in the previous phases is used to identify possibilities for improvement. After evaluation, the cycle starts again with a new strategic analysis, using the insights gained in the preceding cycle. This is reflected in the model, proposed by Rosenbloom and Burgelman (1989), which sees the process of strategy making essentially as a social learning process based on knowledge about which actions have led to success or failure in the past (e.g. Selznick, 1957; Burgelman, 1988). In their model, technology strategy emerges from organizational capabilities shaped by the generative forces of the firm's strategic behavior and the evolution of the technological environment, and by the integrative mechanisms of the firm's organizational context and the environment of the industry in which it operates. This model forms the basis for the dynamic model underlying the longitudinal survey discussed in chapter 4.

The industrial organization as well as the competence perspective on strategy recognizes innovation as an important factor for economic success. Porter (1985) observes that innovation is among the most prominent factors that determine the rules of competition because innovations may affect each of the five forces discussed in Chapter 2. Hence, an active innovation strategy may serve as an effective tool for pursuing a firm's corporate strategy. Burgelman and Rosenbloom (1997) argues that innovation is as important to the firm's long-term survival as its financial and human resources. Therefore, the management of innovation should be regarded as a basic business function and companies should consequently develop an innovation strategy analogous to financial and human resource strategies. This requires that the management is able to assess the firm's innovative competencies and capabilities, identify how they may be improved and allocate resources to leverage the firm's innovative performance.

However, the strategic role of innovation differs according to the strategic perspective chosen. In the industrial organization perspective, the structure of the industry is the starting point. The market leads and the firm has to adapt. The influence of the company's strategic actions on the environment is considered to be limited and the strategic choices a company makes are primarily a reaction to its analysis and knowledge of the industry structure. In this perspective, innovation is often seen as a defensive instrument allowing the creation of barriers against new intruders to protect the firm's market position. But the industrial organization theory also has a major role in understanding the part played by innovation in chains or networks of companies. According to the strategic forces model, the most important factor at supply chain level is the balance of power between supplier and buyer. Depending on the type of market and chain, innovation can offer advantages either to suppliers or buyers. Based on a study of 114 suppliers in the automotive industry, Kamath and Liker (1990) underlined the importance of this relationship of dependency. The most dependent suppliers were prepared to invest (sometimes large sums) in innovation if they knew that the investment was desired by the customer, even if it was not profitable for them from a purely economic perspective.

In the competence perspective, the company's ability to influence the market is considered to be of crucial importance because the company is primarily seen as a bundle of unique resources which should be exploited in the market as a plus versus competition. These unique resources are the starting point for a firm's innovation strategy. Innovative performance then depends on the availability of input factors that are critical to the innovation process (e.g. Cooper and Kleinschmidt, 1995; Gopalakrishnan, 2000; Hadjimanolis, 2000). These input factors can be tangible or intangible, and human, physical, technological or reputational (Hadjimanolis, 2000). Tangibles are the physical assets that enable the organization to undertake innovation-related activities, such as finance, human resources and facilities. They will primarily be of a property-based nature (see Miller and Shamsie, 1996). Although the need for money to undertake innovation projects is intuitively logical, the availability of financial resources has also been 'officially' identified as one of the most prominent bottle-necks in innovation (Ernst, 2002; Hüsig and Kohn, 2003). This has not only to do with the total research budget but also with the mechanism to make it available: from central funding to tying a large portion of the research budget to business division contracts (Buderi, 2000). Intangibles, on the other hand, are the non-physical assets that fulfill the same role, such as the knowledge available within the organization (in the form of patents, copyrights, trade secrets etc.), the brand names owned by the firm and the organization's image or reputation.

The importance of integrating the two approaches to strategy when analyzing innovation strategy is indicated by Kline and Rosenberg (1986), who pointed out that success in innovation is the result of the effective combination of factors proposed by the industrial organization as well as the competence perspective, and the careful tuning of internal competencies and capabilities to the technological challenges and business opportunities. Winners in the global market-place are those firms that can demonstrate a timely proactive response to changing situations and rapid and flexible product innovation, coupled with the management capability to effectively coordinate and redeploy internal and external competencies. Companies that have been successful over long periods of time have developed capabilities that are quite distinct from those of their competitors and are not easily replicable. The strategies of such companies cannot be classified simply in terms of differentiation or cost leadership; they combine both. The ability to maintain a uniqueness that is salient in the market- place, however, implies continuous innovation and being alert to what competitors are doing and should not be confused with an inward-looking orientation.

How the two perspectives are interlinked was also demonstrated by McGrath (1995) who stated that technology-based companies operate under turbulent conditions due to the fact that both their markets and products are technology driven. This creates unique strategic challenges since they have to:
- constantly build new markets;
- manage short and rapidly changing product life cycles;
- harness emerging technology; and
- adapt to collapsing markets.

In order to survive and grow, they have to meet these challenges not one by one, but all at the same time.

3.8 Analyzing strategic alignment in the case of innovation

We concluded in Section 2.8 that from an industrial organization perspective the proposed factors will be found in the business environment, industry and market in which a company operates. Empirical research has focused on three key dimensions of the business environment: environmental uncertainty, dynamism and complexity (Dess and Beard, 1984). As Peng and York (2000) indicate, the business environment includes more than only the economic players (competitors, buyers and suppliers). It also includes institutional and cultural dimensions. Luo and Peng (1999) classified the business environmental factors as follows. Environmental uncertainty, what they call environmental hostility, refers to the level of deterrence and the possible impact on the firm's performance of the economic players, the legal situation, and socio-cultural groups. Environmental dynamism refers to the level of predictability of the actions of these groups or factors, whereas environmental complexity refers to the diversity of external factors, i.e. how many factors the firm has to cope with simultaneously, and heterogeneity of these factors. According to Volberda (1992), the business environment can best be typified by the level of market dynamism because dynamism describes the degree to which elements of the environment in which an organization operates either remain basically the same over time or are in a continual process of change. For that reason we have chosen the term market dynamism to reflect the combined influence of environmental uncertainty, dynamism and complexity as defined by Dess and Beard (1984).

An important internal factor for technology-based companies operate is the core technology that underlies their innovative activities, such as biomedical and pharmacological research in the case of pharmaceutical companies, micro-processor technologies for electronic firms and aeronautics in the case of the aircraft industry. These technologies not only differ with respect to the complexity and number of elements needed to introduce successful innovations but, more importantly, in the interdependence of these elements. According to Volberda (1992), it is primarily this last aspect that determines the level of complexity. From this, we conclude that the main factors that may be expected to affect the alignment of innovation to business from the industrial perspective will be market dynamism and the technological complexity of the environment in which a company operates. Investigating the effect of these factors can best be done by studying the content of the innovation strategy under different market and technology conditions, along the line of the strategy formulation-implementation approach of alignment (see Table 2.3).

In Chapter 2, we concluded that the competence perspective proposes the firm's own resources, competencies and capabilities as the key factors to study in questions of strategy. As we set out in Section 3.3, a firm's most relevant competencies and capabilities in the case of innovation are its ability to create, acquire and exploit new knowledge, and its ability to create and maintain effective links within the innovation function (cross-functional communication) and with its business environment (market orientation). In addition to these factors, the availability of R&D resources (R&D funding structure) and management tools are identified as important resources in this perspective. Finally, the competence perspective stresses the ability of a company to learn. In the case of innovation this means, that the effect of feedback should be added as a potential influential factor for the process of aligning innovation to business. Investigating how these factors affect strategic alignment requires studying the process of implementing the innovation strategy by the internal network that constitutes the innovation function of the company. This means using the network-process approach to alignment (see Table 2.3). This approach pays tribute to the fact that, especially in the case of innovation, strategic alignment is not an event but a process of continuous adaptation and change because strategic choices made by the firm will, as Henderson and Venkatraman (1993), observed, inevitably evoke counteractions (e.g. imitation, own innovations) by its major competitors that will necessitate a subsequent response.

3.9 Concluding remarks

In this chapter we applied the insights of how the phenomenon of strategic alignment can be linked to innovation. We conclude that, from an industrial organization point of view, it can be expected that the main factors affecting the alignment of innovation to business will be market dynamism and the technology complexity characterizing the (inter-)national industrial environment in which a company operates. We also conclude that investigating the effect of these factors can best be done by studying the content of the innovation strategy in technology-based companies that are operating under different market and technology conditions along the line of the strategy formulation-implementation approach of alignment.

From a competence perspective, we conclude that the main factors affecting the strategic alignment of innovation to business will be found in a firm's innovation resources, innovation competencies and capabilities of the firm. Investigating how these factors affect strategic alignment requires studying the process of implementing the innovation strategy by the internal network that constitutes the innovation function of a company. This means using the network-process approach that acknowledges the inherently dynamic nature of aligning innovation to business strategy.

4. Research design

In this chapter, we discuss how we relate the insights on the methods to analyze the phenomenon of strategic alignment in the domain of innovation to the design of an empirical investigation. We do so by presenting the conceptual framework and selecting a sound methodology that will enable us to arrive at scientifically accountable answers to our research questions. In Section 4.1, we start by developing a general conceptual framework underlying the study as a whole. In Section 4.2, we formulate our basic methodological starting points and present the overall research design which consists of two consecutive and complementary studies: the cross-industry study and the longitudinal study. In Section 4.3, we focus on the cross-industry study by means of the integrated formulation-implementation approach, as described by Venkatraman and Camillus (1984). We first propose the product generation life cycle as the main indicator for assessing the combined effect of market dynamism and technology complexity. The conceptual framework of the cross-industry study is then presented as are the propositions related to the internal and external fit of the innovation strategy. Finally, the operationalizations of the research variables, the expected relationships and the methods of data collection and analysis are outlined. In the longitudinal study design, presented in Section 4.4, the focus shifts to the dynamic character of strategic alignment by investigating how the internal and external fit of the innovation strategy develops over time as perceived by the R&D staff on the one hand and the business unit and headquarter's managers on the other, using the integrated network-process approach proposed by Venkatraman and Camillus (1984). This section provides the conceptual framework, the set of propositions to be tested, the operationalizations of the research variables, and the methods of data collection and analysis for the longitudinal study. The chapter ends with some concluding remarks in Section 4.5.

4.1 The general conceptual framework

In Section 3.7, we concluded that the main factors that affect the formulation, implementation and execution of an innovation strategy and its alignment to business from the industrial organization point of view are the market and technology forces that characterize the industrial environment in which a firm operates, and that these factors can best be investigated using a strategy formulation-implementation approach. We also concluded, that from the competence perspective the main factors affecting strategic alignment of innovation to business are found in the innovation resources, competencies and capabilities of a firm and that these factors can best be investigated using a process-network approach.

In Figure 4.1, these factors are combined in the general conceptual framework underlying the present study. The technology-based firm is shown as a bundle of resources, competencies and capabilities that concentrates on product and process innovation in order to be first on the market. The actors in the technology-based firm responsible for crafting the innovation strategy and its internal and external alignment are modeled as the firm's innovation function. It includes the Headquarters (HQ), Corporate R&D (R&D) and Business Units (BUs)-

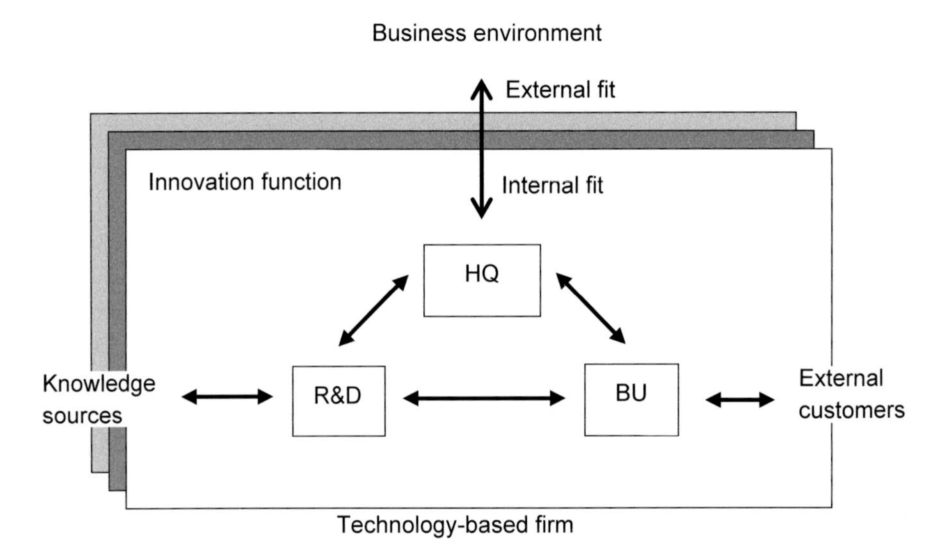

Figure 4.1. General conceptual framework.

triangle. We consider HQ, R&D and the BUs as the nodes, and the communication (e.g. project plans, project results and services) and the governance structure (decision rights, accountability and coordination) as the linkages in this internal network. Within this triangle, the innovation strategy is formulated, implemented, executed and evaluated. The framework shows that the firm as such is also part of an external network, with the internal nodes having external linkages with the business environment, i.e. the link of corporate R&D to external knowledge sources such as universities, suppliers and buyers, and the link of the BUs to their external customers. Where external fit refers to proactively attuning the innovation function to the business environment, internal fit refers to proactively attuning the innovation resources, competencies and capabilities in the HQ-R&D-BU triangle. Finally, strategic alignment is defined as a function of the achieved level of internal and external fit.

4.2 Overall research design

In Section 3.7 we concluded that investigating the effect of the factors, proposed by the industrial organization theory can best be done by studying the content of the innovation strategy in technology-based companies, operating under different market and technology conditions. However investigating the factors proposed by the competence perspective might better be done by studying the process of implementing the innovation strategy by the internal network that constitutes the innovation function of the company. This necessitates a dual methodology capable of capturing cross-industry differences as well as dynamic processes. From a methodological perspective, using a dual methodology offers the opportunity for complementary and synergistic data gathering and analysis. This is expected to create synergy and enhance the internal and external validity of the empirical studies. The first empirical study was set up to address the first research question.

RQ1. What is the effect of the industry 'clockspeed' on the strategic alignment of innovation to business?

In Section 3.7 we concluded that investigating the effect of these factors can best be done by studying the content of the innovation strategy in technology-based companies operating under different market and technology conditions, i.e. along the line of the strategy formulation-implementation approach of alignment. It was therefore decided to conduct a multiple case study of technology-based companies across industries that differed according to their industry 'clockspeed'. For this study an inductive research strategy was chosen (Eisenhardt 1989), with a priori specification of the research variables: the technology-based firm, its innovation strategy and its business environment. The firm characteristic of culture and the industry characteristics of place in the technology life cycle and length of the PGLC were used as possibly important a priori dimensions, to distinguish within-group similarities coupled with inter-group differences. Of these, the PGLC proved to be the most powerful one to distinguish a pattern between the investigated firms, that is, by classifying the firms into two groups, one with PGLCs shorter than 6 years, and one with PGLCs longer than 6 years, a consistent pattern of differences between the two groups could be observed. For this reason the PGLC was selected as the moderating variable to distinguish across industries. The second part of the research was then of a deductive nature, by linking the PGLC concept to the strategy literature and formulating propositions that might explain the mechanisms underlying the differences in innovation strategy related to different PGLC length.

Yin (1989) argued that the logic underlying a multiple case study approach is similar to that guiding multiple experiments and that each case should be selected so that it 'either (a) predicts similar results (a literal replication), or (b) produces contrary results but for predictable reasons (a theoretical replication)'. The latter is the case in the cross-industry survey in which ten technology-based companies were selected so as to deliberately vary the market and technology forces, as reflected in the different lengths of the product generation life cycle (PGLC, see Section 4.3.1).

Theoretical replication requires that the phenomenon being studied be defined by some characteristics common to all the research situations, apart from the differences to be investigated (Yin 1989). In the cross-industry study, we tried to control for irrelevant sources of variance originating within the firm by selecting only large, technology-based companies that were each among the top three in their respective industries. The ten companies included in the present study match were all high-performing, technology-based firms with high inputs (in terms of their R&D expenditures related to their industrial sector) and with a corporate R&D facility.

The second empirical study, the longitudinal study, was set up to address research question 2.

RQ2. How can strategic alignment of innovation to business be achieved and maintained over time?

To address RQ2, a longitudinal study with a duration of six years was conducted in one of the companies included in the cross-industry survey, a multinational supplier of industrial components. From 1997 to 2003, the strategic alignment of the corporate R&D center and their internal customers in the business units was studied by means of a biannual survey questionnaire that covered the R&D competencies and capabilities. The biannual survey fulfils a number of functions. First of all, such surveys are designed to gain an understanding of how the internal factors affect strategic alignment. Secondly, since all actors in the innovation function (HQ, R&D, and the BUs, see Figure 4.1) are included, the outcome enable us to analyze differences in the pattern of answers given by the respondents, and by doing so determine perception gaps between the different actors. In the course of the investigation period, the surveys functioned as a feedback mechanism as well. The results of each survey were reported to the HQ-R&D-BU triangle, and if necessary the management responsible took action.

4.3 The cross-industry study

As concluded in Section 3.7, the main forces that technology-based companies are facing in their business environment when formulating and implementing their innovation strategies are market dynamism and technological complexity. To investigate the effect of the business environment on the strategic alignment of innovation to business in the cross-industry survey, we first have to find an indicator that can be used to assess the combined effect of these forces. We propose that such an indicator could be found in the concept of the industry 'clockspeed' (Brown and Eisenhardt, 1998), indicated by the length between the subsequent product generations, further referred to as the Product Generation Life Cycle (PGLC) , which is discussed below.

4.3.1 The product generation life cycle

The product generation life cycle (PGLC) is build on the well-known concept of the product life cycle (PLC). Bayus (1994) defines the product life cycle as the evolution of unit sales over the entire lifetime of a product. The product life cycle (e.g. Cox, 1967, Levitt, 1965, and Polli and Cook, 1969, Moore, 1995) has four stages: introduction (an initial period of slow sales growth), growth (a period of rapid growth in sales), maturity (a period in which sales level off and are relatively stable), and decline (when sales drop off).

Following Maidique and Zirger (1985), we argue that the PGLC-concept, the sum of the product life cycles of all related products belonging to one product generation (what they refer to as the product family) is a far superior unit of analysis when studying innovation management than the PLC, because the PGLC incorporates the interrelationship between products. We add to this argument that the PLC of a single product can be severely influenced by trends and regional preferences. In addition, the PGLC-concept links up with the common industry practice since the 1970s of technology-based firms to design 'product platforms', groups of products aimed at different market segments or customer groups, but using the same basic product architecture to reduce the complexity of new product development. This allows them to rapidly introduce variations while delaying periodic changes in the architecture for as long as possible (Meyer and Utterback, 1993). In this book, we use the following definition for the product generation life cycle.

The product generation life cycle is the time span between the time that the first product of a product generation is delivered to the customer and the time when the total volume of production of this product is just 10% of its maximum.

We chose for 10% of the peak sales volume as the end point of the PGLC following Fisher and Pry (1971) who used this cut-off point to measure individual product substitution. In practice, a level of 10% is easier to measure than the point in time at which no sales occur anymore, for instance, some sales can go on in niche markets long after the main products are withdrawn from the market.

The first question we have to answer is whether the PGLC can be used as an indicator to reflect the differences in market and technology forces among industries. Sanderson and Uzumeri (1997) indicated that this is indeed possible. They observed in case study research that interactions between variety and rate of change in product families seem to be determined by the forces of technological change and market demand. To be useful as an indicator, the PGLC must fulfill the following requirements.

- It must show clear differences across industries.
- It should reflect the market as well as the technological forces present in the firm's business environment.
- The nature of these relationships must be clear and predictable.

We elaborate on these requirements below.

Differences across industries

As Williams (1998) indicated, huge differences in the average PGLC can be observed across industries, ranging from less than a year to over 20 years. Fine (1998) and Brown and Eisenhardt (1998) refer to these differences by introducing the concept of the industry 'clockspeed'. There are industries, like the personal computer, the semiconductor and the mobile phone industries, where product generations follow each other at a dazzling pace. Fine (1998) refers to these industries as 'fruit fly industries', because of their very rapid evolutionary rate. Other technology-based industries, such as heavy electrical equipment, energy, chemicals, including fine chemicals and pharmaceutics, as well as commercial aircraft, satellite and space launcher industries are characterized by relatively long PGLCs. These industries in general have high entrance barriers, for example regarding R&D intensity, marketing quality and/or capital investment. For instance, the pharmaceutical industry is characterized by a long and expensive R&D process and huge marketing costs. It takes about a decade to get a new drug on the market and developing costs have increased dramatically in the last two decades to about a US$ 250 m. Such an expensive endeavor can only be undertaken if there is at least some certainty that the new drug will remain on the market for a long period of time. A defensive patenting strategy on the active compound, combined with a worldwide marketing and sales effort, will help to ensure this. In other industries with comparable challenges, we see the same pattern. The design of a Boeing 747 today is basically the same as when it was first introduced in 1968, although much has changed in the technology within the airplane, especially in the electronic systems. These observations suggest that the product generation life cycle is indeed a fundamental contrasting variable between industries.

Market and technology forces

Does the length of the product generation life cycle reflect market as well as technology forces in the business environment in which a company operates and is the nature of this relationship clear and predictable?

Of course, technological forces and market forces are both present in all industries, but their relative strengths and thereby their combined effects may vary across industries. In typical long life cycle industries (further referred to as LLCIs), like the aircraft and pharmaceutical industries, companies are generally confronted with high technology complexity and/or low (or even zero) defects tolerance. This has to do with the fact that no one wants to die in a plane crash or suffer the side effects of using an untested medicine. High technology complexity results in high uncertainty in the 'fuzzy front end', which will necessitate a higher research focus resulting in an elongation of the R&D process, while low (or even zero) defects tolerance will lead to a high level of quality control that will cause an elongation of the development and engineering processes. Both will lead to a longer time span between the subsequent market introductions, and thus to an elongation of the PGLC.

Companies in industries with relatively short life cycles for the main products (further referred to as SLCIs) are typically confronted with a high level of market dynamism and competition leading to pressure to speed up the R&D process in order to shorten the time-to-market. This will mean a shorter time span between the subsequent market introductions, and to a shortening of the PGLC. In short: if technology forces prevail due to technological complexity or zero defect tolerance, the length of the PGLC will tend to longer. When market forces prevail, PGLCs tend to shorten, as was observed in the computer industry in the past decade. From these observations, we propose that the length of the PGLC reflects the combined effect of technology and market forces in a given industry, and that the nature of the relationship between the length of the PGLC and the market and technological forces is clear and predictable. The length of the PGLC can therefore be used as an indicator for a cross-industry comparison of technology-based companies.

4.3.2 Conceptual framework

Figure 4.2 shows the main concepts investigated in the cross-industry study: the firm characteristics of the technology-based prospector company, its innovation characteristics, the strategic alignment of innovation to business, the length of the PGLC of its main products, and the methods it applies to improve strategic alignment.

A technology-based company is characterized by its resources, culture and performance. The R&D strategy is characterized by the second and third phases of the process of crafting an innovation strategy from strategic analysis to strategy evaluation, the formulation phase, and the implementation and execution phases. According to Caloghirou *et al.* (2004) important elements that characterize these phases are a firm's investment in R&D (discovery capability), in appropriating R&D outcomes (patenting), and in human resources training

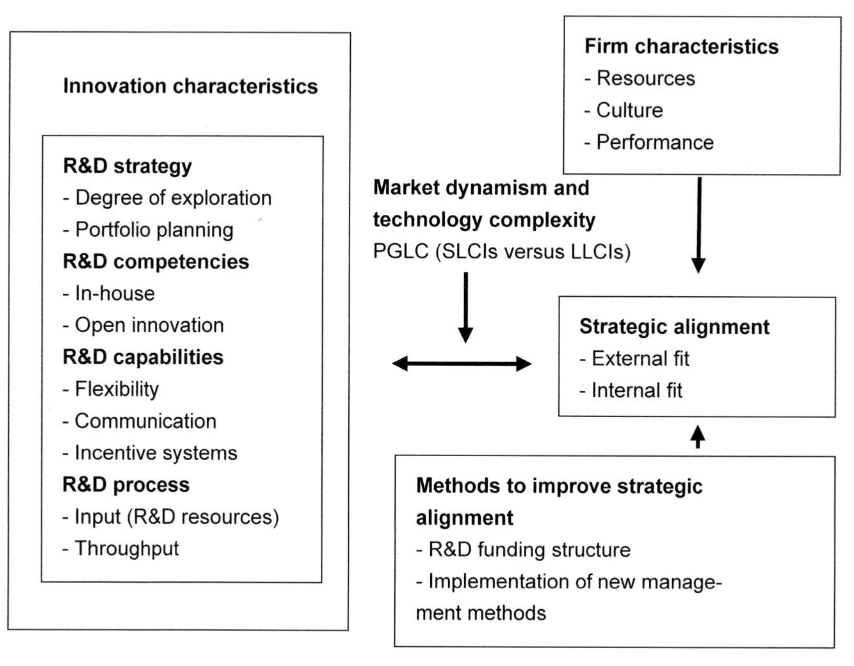

Figure 4.2. Conceptual framework of the cross-industry study.

(learning and knowledge transfer capability). The formulation of an innovation strategy is considered to reflect the innovative orientation of the firm. It will either be more exploration or more exploitation-oriented. The implementation of the innovation strategy creates the conditions for an effective process of knowledge absorption, the creation of new knowledge and knowledge transfer to be utilized in the conversion of knowledge into new value. It involves the allocation of resources, and the planning and organizing of the competencies (in-house technical skills and know-how, and open innovation) and capabilities (flexibility, communication and incentive systems). The execution of the innovation strategy is reflected in the innovation process: input (R&D resources), throughput and the respective outputs in the research, development and commercialization phases. Strategic alignment is conceptualized as the combination of the level of fit to the external (business) environment and the level of fit to the firm's internal resources, competencies and capabilities. The strategic alignment of innovation to business is considered to be moderated by the length of the PGLC.

4.3.3 Operationalization of the research variables

To test the conceptual framework (Figure 4.2) empirically, the concepts are split into research variables that are operationalized by providing operational definitions and indicators to produce measurable items (see Table 4.1).

The exact operationalization of the different research variables in Questionnaires I and II are presented in Appendices B and C respectively. Questionnaire I requests quantitative and factual information regarding the company as a whole (e.g. sales volume, profitability, and market share of the different BUs), and specific information about the corporate R&D center (e.g. R&D budget, R&D personnel, number of patents and R&D management systems). Questionnaire II asked for personal perceptions about the quality of the firm's R&D competencies and capabilities and the level of strategic alignment, using Likert seven-point scales. This questionnaire was presented to a number of R&D department heads and/or R&D program managers in each company. In Section 4.3.6 the questionnaires are discussed in more detail.

Table 4.1 starts by introducing the PGLC, which is used as an indicator for the firm's business environment (see Section 4.3.1), and a baseline description of the technology-based firm, including the firm's R&D resources, culture (Anglo-American or continental European headquarters), and its performance in terms of sales volume and operating profit margin.

Strategic alignment is operationalized as the external fit with the business environment, for example in terms of monitoring market and technology trends, and innovation opportunities, and the internal fit of innovation to business, for example in terms of the R&D objectives set in line with business plans and business risk evaluation in the R&D portfolio. *R&D strategy* is assessed as the degree of exploration and is measured quantitatively on the R&D input and R&D output sides. On the R&D output side, the revenue contribution of new

Table 4.1. Operationalization of the research variables, measures (question numbers, see Appendices A, B and C), and proposed differences between short life cycle industries (SLCIs) and long life cycle industries (LLCIs).

Research variables	Operational definitions and indicators	Measures	SLCIs	LLCIs
Business environment				
Market dynamism and technological complexity	SLCIs: PGLC < 6 years; LLCIs: PGLC > 6 years; See R&D performance: sales revenues from new products	B3, B4, B6, B7		
Baseline description of the technology-based firm				
Resources	Total number of employees, and per division	$B1_{2}$, B2, B13, B15		
R&D resources	R&D expenditures, R&D as percentage of sales	$B1_{1,6,7}$		
Culture	Anglo-American versus continental European headquarters			
Performance	Sales volume and operating profit margins	$B1_{1,4}$		
Strategic alignment				
External fit	Market & technology trends important strategy inputs; R&D portfolio based on technology and market vision; monitoring of innovation opportunities; market-technology roadmaps	A1, C1-3, 5	+	
Internal fit	R&D projects meeting business goals; R&D and BU viewpoints on meeting business goals; R&D objectives in line with BU business plans; cost drivers, capital constraints, R&D, manufacturing, business risks taken into account	B20, 30_{1-4}, 31, C9, 20, 25, 28	+	
R&D strategy				
R&D performance	Sales revenue from new products	B4, 6, 7	+	
	Number of patents per US$ 10 m	B5		+
Degree of exploration	Research as percentage of total R&D, PhDs as percentage of total R&D staff	A2, B8, B9-11, C18		+
	Funding for basic research easy to get; senior management involved in early development	C26, C35, C36		+
	Speed emphasized over budget; importance of value engineering	C10, C11, C18	+	

Table 4.1. Continued.

Research variables	Operational definitions and indicators	Measures	SLCIs	LLCIs
R&D competencies				
In-house	Core technology definition, R&D competencies monitoring and exploitation	A3.1, $B13_1$, C4, C6, C8	+	
Open innovation	Collaboration with suppliers and customers	A3.2, $B13_{2-4}$, B14, B15, $B22_{4-6}$, C7, C30, C37		
	Collaboration with research firms and knowledge institutions (RTIs)	A3.2, C34		+
R&D capabilities				
Flexibility: timeliness	Average project cycle time; percentage of cycle time reduction; time to move to new research field, speed of appointment and purchase, and bureaucratic constraints	A4, A4.1, B10, B18, B19, B35, B36, C40	+	
Flexibility: responsiveness	Ease of incorporation of BU requests in the corporate R&D portfolio	B37	+	
Internal communication	Cross-functional (ICT) communication, communication with the BUs (reporting, meetings), marketing & sales, and network communication	A4.2, $B22_{1-3}$, B24, B25, C1, C12-15, C24, C30, C31	+	
External communication	Communication with external customers	$A4.3_1$, $B22_{4,5}$, B23, B32, C16, C33	+	
	Conference communication, communication with experts	$A4.3_2$		+
Incentive systems	Monetary incentives, result-based competition, recognition, recruitment, learning and education	A4.4, B39, B40		
R&D process				
R&D throughput	Percentage of time spent on different phases of R&D; stage gate review process	A5, B16-17, B21, B22, B26-29, B33, B34, C23		

Table 4.1. Continued.

Research variables	Operational definitions and indicators	Measures	SLCIs	LLCIs
Methods to improve strategic alignment				
R&D funding structure	Percentage of R&D funding from headquarters versus BU funding	A6.1, B12		
Implementation of new management methods	BU forced to generate sales from innovative products; R&D projects evaluated in terms of alignment to business; R&D projects prioritized based on customer value; QFD, DFX, virtual development (modeling and simulation)	A6.2, C4, C17-19, C22, C38, C39		
Communication improvement methods	Cross-functional participation in R&D planning; R&D project prioritization by BUs; R&D project parameters discussed with BUs; R&D participation in BU business plan formulation; staff exchange	B38, C19, C27		

+ indicates a higher proposed level

products and the number of patent applications are measured, and on the input side the percentage of the total R&D budget that is spent on fundamental technology development and the number of staff with a PhD as a percentage of the total R&D staff are assessed. The qualitative assessments include the importance of R&D speed versus budget, the importance of value engineering, and the attention paid by senior management to early development. *R&D competencies* include the in-house R&D competencies as well as the external R&D competencies, either cooperation with chain partners or with research firms and Research and Technology Institutes (RTIs). *R&D capabilities* directed to shorten the time-to-market include timeliness, for example methods to reduce R&D project cycle times and time to move to a new research area, and responsiveness: the time and effort it takes to include a BU request in the R&D portfolio. In addition, the level of internal cross-functional communication with manufacturing, marketing and sales, and external communication with end users and incentive systems to stimulate innovation, are measured. Of the *R&D process*, the time spent on the different phases of basic and applied research and development is measured. Whether a stage gate review process is present is also investigated. Finally, a number of *managerial methods to improve strategic alignment* are discussed, such as BU funding and the participation of BUs in R&D planning and staff exchanges.

4.3.4 Propositions

This sub-section provides the propositions and expected differences between SLCIs and LLCIs based on these propositions (see Table 4.1) that were tested empirically in the cross-industry study.

Strategic alignment

The first proposition regards the strategic alignment situation in SLCIs and LLCIs.

P(C)1. The level of strategic alignment between innovation and business will be higher in SLCIs than in LLCIs, based on the closer market proximity in SLCIs.

The conceptual framework indicates that we expect that differences in time horizon, caused by differences in PGLC lengths will be reflected in the strategic alignment of innovation to business between companies in SLCIs and LLCIs. We expect that strategic alignment will be more problematic in LLCIs than in SLCIs. In SLCIs the relatively short time span between the subsequent product generations implies that the time horizon for R&D to develop new products is also short.

This suggests that R&D is relatively close to the market, which makes it easier to keep R&D and business strategies aligned, both in terms of internal and external fit. The external fit, as reflected in the importance attached to market and technology trends as strategic inputs but also in the use of tools such as market-technology roadmaps, is expected to be better in SLCIs. The internal fit, reflected in the extent to which R&D priorities are set in line with business plans and R&D projects meet business goals, is expected to be better in SLCIs because of more intensive internal communication and the shorter time-span between the start of the R&D process and the final market introduction. R&D strategy.

The second proposition regards the R&D strategy.

P(C)2. The relatively high market dynamism in SLCIs will lead to a more exploitation-oriented R&D strategy. The relatively high technology complexity in LLCIs will lead to a more exploration-oriented R&D strategy.

As we have seen in Section 4.3.1, LLCIs are characterized by a relatively high level of technological complexity. This high level will have a number of consequences for the R&D process. It means that more time and effort will have to be put in the fuzzy front end of the R&D process and more R&D efforts will be needed to coordinate the development of the different components or sub-systems. The low or even zero defect tolerance will result in longer development and testing procedures to ensure strict adherence to technical specifications or proof of absence of harmful side-effects, for example in pharmaceutical firms. This will lead to a longer time-span between the different product generations and a higher R&D investment per new product. It is expected that this will result in an R&D process that is more oriented towards fundamental research and technology development. It is therefore anticipated that the R&D strategy in LLCIs will tend to be more exploration-oriented, reflected in a strong emphasis on knowledge creation and acquisition but also on knowledge protection to ensure that costly investments will finally pay off.

On the other hand, in SLCIs the high market pressure will result in high in-company pressure on R&D to come up with new products at a rapid pace. To meet these demands, R&D will be more focused on incremental innovation, on responsiveness to market priorities and the timely delivery of project results, and on fast knowledge transfer to BUs. This will be reflected in R&D performance. The higher market pressure in SLCIs will result in an inherently higher percentage of sales derived from new products, whereas the higher emphasis on knowledge creation and protection in LLCIs will be reflected in a higher percentage of patents per R&D investment. In addition, the higher degree of exploration in LLCIs will be reflected in a greater part of the total R&D budget being allocated to research, and a staff composition of the corporate R&D center will have a higher percentage of personnel with a PhD. The more longer-term technology outlook of LLCIs will be reflected in a higher level of senior management involvement in the early phases of the development process. The high market pressure in SLCI will lead to a greater emphasis on value engineering and speed as the most important objectives in the R&D process (see Table 4.1).

R&D competencies

Regarding R&D competencies, we expect to find a different orientation towards open innovation between SLCIs and LLCIs.

P(C)3. The more exploitation-oriented R&D strategy in SLCIs will lead to R&D competencies being more focused on in-house knowledge. The more exploration-oriented R&D strategy in LLCIs will lead to R&D competencies being more focused on open innovation through collaboration with external knowledge sources.

The stronger focus on knowledge exploitation by SLCIs will force them to pay more attention to in-house R&D competencies in order to get new products on the market as fast as possible. We expect that the more exploration-oriented strategy of LLCIs will result in more emphasis on research collaboration with research firms and knowledge institutions to further develop the technology knowledge base of a company.

R&D capabilities

P(C)4. The more exploitation-oriented R&D strategy in SLCIs will lead to R&D capabilities being more directed at increasing customer orientation and speed to market. The more exploration-

oriented R&D strategy in LLCIs will lead to R&D capabilities being more directed towards external knowledge acquisition.

Regarding R&D capabilities, we expect that the high market pressure in SLCIs will create high in-company pressure on R&D to reduce its throughput times. This will be reflected in a stricter stage gate review system and a higher percentage of realized reduction of R&D project cycle time in the preceding three years. Furthermore, we expect that the relatively high market dynamism will pressurize SLCIs into paying a great deal of extra attention to R&D flexibility and efficient communication to ensure rapid internal knowledge transfer, which will be reflected in a higher level of responsiveness (ease of incorporation of BU requests into the R&D portfolio), a higher level of timeliness (e.g. time-to-market, time to move to a new research field, less bureaucratic constraints), better internal communication (cross-functional communication) and more intensive communication with external customers to improve products and services. In contrast to this, we expect that a higher degree of exploration will induce LLCIs to place more emphasis on external scientific communication, which will be reflected in a higher level of conference attendance and collaboration with external experts.

Regarding incentive systems, we do not expect to find differences among SLCI and LLCI companies, since they are all technology leaders in their respective industries, and are trying to find and keep the best staff available. There is no reason why this would be affected by the differences in time horizon and the underlying factors of technology complexity and market dynamism. Regarding open innovation with chain partners (buyers and suppliers), the organization of the R&D process and the methods to improve strategic alignment, we expect to find differences between SLCIs and LLCIs but we cannot predict the kind of differences upfront.

4.3.5 Data collection

Data were collected in 1997 and 1998. Fifteen eligible companies were identified using the UK R&D Scoreboard (www.innovation.gov.uk) that lists the 1,000 global companies with the highest R&D investments. The selected companies come from different industries and their main products all have a different length of PGLC. However they share the fact that they are all leading multinational technology-based prospector companies and have a high level of worldwide R&D expenditure relative to their industrial sector. In each company, the chief technology officer (CTO) or the Director of the Corporate R&D center was identified. They were sent an introductory letter explaining the objectives of this study and the research questionnaires were exposed. As an incentive to participate an individualized report was offered to each company, showing their results compared to the general findings. The letter was followed by a telephone call two weeks later to ask for cooperation and to answer any questions they might have. After approval, site visits were planned. Structured interviews were held with the CTO or the Director of the Corporate R&D center and, where necessary, with technology directors. The number of interviews depended on the complexity of the R&D

situation in terms of the number of business areas and the firm's level of centralization. In these interviews the strategy, structure and organization of R&D were discussed (see Appendix A). To avoid any misunderstanding, the transcripts of the interviews were sent to the interviewees for their approval. One copy of Questionnaire I (see Appendix B) was filled out per company by the CTO or the Director of the Corporate R&D center, and where possible the answers were checked using publicly available data. Questionnaire II (Appendix C) was presented to a number of R&D department heads and/or R&D program managers to ensure a more balanced representation of the items at issue. Appendix E provides a glossary of the terms used in the research questionnaires.

4.3.6 Development of the survey questionnaires and methods of data analysis

To put together Questionnaires I and II, approximately 200 questions were initially collected based on in-depth expert enquiry and the work of Roberts (1995) who conducted an extensive study of the 95 large technology-based companies in the United States, Western Europe and Japan in the early 90s. These 200 questions were subjected to two stages of data refinement using a panel of five experts from industry and academia to ensure content validity (see Section 4.4.3). The first stage focused on condensing the questionnaires by retaining only those items capable of discriminating between respondents, examining the dimensionality of the scales and establishing the reliability of its components. The second stage was primarily confirmatory in nature and involved re-evaluating the condensed scales' dimensionality and reliability by retesting them. The questionnaire used seven-point Likert scales, ranging from 'strongly disagree' (1) to 'strongly agree' (7), with no verbal labels for the intermediate scale points. Several items were negatively worded to reduce response tendencies by respondents (Cooper and Emory, 1995). These items were reverse-scored for use in the analyses in order to ensure that a higher assessment in all cases reflected a more positive judgment of the question at issue. The use of structured interviews, a quantitative questionnaire and a qualitative questionnaire also provided for triangulation because different types of data were gathered which could be used as a crosscheck and to draw on the particular and different strengths of the various data collection methods (see Pettigrew, 1990). To avoid respondents providing too positive a picture, the answers in the structured interviews and those given in Questionnaire I were cross-checked with the information provided in public documents, i.e. in annual reports. Data analysis consisted of comparing group means (SLCIs versus LLCIs). For clarity of presentation, we use parametric methods, average, standard deviation and t-tests. We chose for one-tailed t-tests because propositions were developed concerning the direction of expected relationships. Non-parametric statistical analyses using the Mann-Whitney Test did not change the conclusions (see also the footnote on the acceptability of using parametric tests in cases of ordinal scales in Section 4.4.3).

4.4 The longitudinal study

As argued in Section 4.2, investigating how the innovation resources, innovation competencies and capabilities of the firm affect strategic alignment requires studying the process of implementing the innovation strategy by the internal innovation network, which is comprised of corporate headquarters, corporate R&D and business units, in a longitudinal study.

4.4.1 Conceptual framework

Figure 4.3 shows the conceptual framework of the longitudinal study. The essential element in this framework is a comparison of the perceptions of R&D staff, and those of managers at headquarters and in the business units regarding the quality of R&D competencies and capabilities to fulfill a company's long-term and short-term objectives effectively and efficiently. Comparing these perceptions on a regular basis with feedback from the longitudinal survey questionnaires, it was possible to analyze the development of the level of congruence between these perceptions over time. It was assumed that the more congruent the perceptions of the R&D staff and the BU and HQ respondents become, the higher the level of strategic alignment would be. The longitudinal survey questionnaire provided four consecutive measurements over

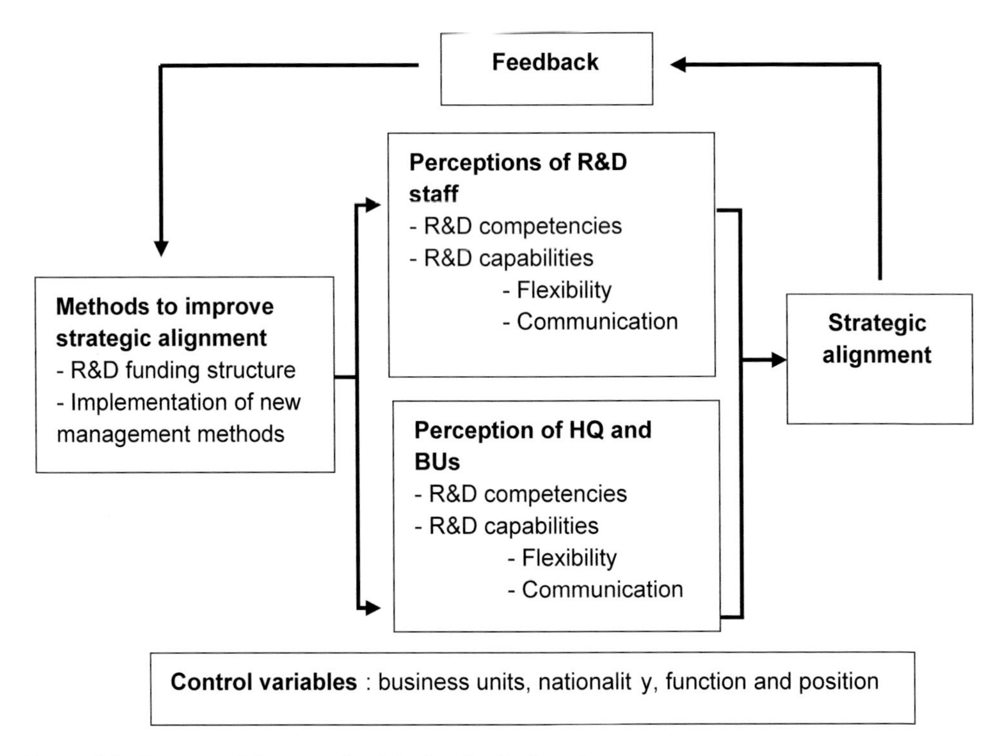

Figure 4.3. Conceptual framework of the longitudinal study.

a period of six years. This had the advantage that the effect of the management methods, such as the introduction of key performance indicators, market-technology roadmaps and changes in R&D funding structure, that were introduced to improve alignment and the learning processes induced by feedback from the survey questionnaire could be part of the investigation. However, it did not allow for the use of a too-detailed questionnaire because the cooperation of many respondents over a longer period of time was only possible, if a questionnaire did not take too much time to fill out. Therefore, for this study only those R&D competencies and capabilities that seemed to be most critical for technology-based firms were selected.

The R&D competencies reflect the technical skills and know-how of the corporate R&D center and consist of the following six technology fields: expanding the company's technology-knowledge base; developing new technology in a product or process area; translating existing technology in new product or process designs; contributing to the improvement of existing product or process designs; developing new product or process tests; and offering new technology for cost reduction. The R&D capabilities reflect the managerial systems of the corporate R&D center, and consist of the level of perceived flexibility and the quality of communication with the BU customers. Flexibility is assessed in two dimensions: responsiveness, which is the time and effort it takes for a business unit to get an important new project incorporated in the R&D portfolio, and timeliness, which reflects the time aspects of executing R&D projects.

4.4.2 Propositions

We elaborate now on the different concepts in the conceptual framework and the propositions derived from it.

Strategic alignment: internal and external fit

The concept of external fit in the longitudinal study was approached from the perception of the members of the groups that form the company's innovation function (HQ, R&D, BUs). It is assessed by measuring how far the R&D activities conducted by the corporate R&D center are aligned to the technology needs of the firm, as well as to the market opportunities.

Inclusion in the longitudinal study of respondents from all three nodes in the HQ-R&D-BU triangle enables us to compare the levels of R&D competencies and capabilities as they are perceived by the R&D staff on the one hand and by the headquarters and BU customers on the other. By following how the gaps between these groups develop over time, these measurements can be used as indicators of changes at the level of internal fit.

R&D competencies and R&D capabilities

As stated earlier, the R&D competencies reflect the technology knowledge base of a firm. Cohen and Levinthal (1990) suggested that the ability of firms to develop new products and processes is determined by their ability to absorb new knowledge and that this ability is, in turn, determined accumulated past experience. Caloghirou *et al.* (2004) therefore argue that the cognitive processes, which constitute the knowledge base of a firm, are cumulative, idiosyncratic and path-dependent in nature. As Kogut and Zander (1992) argued, it is the firm's knowledge base that leads to a set of competencies that enhance the chances of growth and survival. The knowledge base allows the ability to search, recognize and represent a problem as well as to assimilate and use new knowledge for problem solving. It is expected that, as a company's existing knowledge base is better able to absorb new knowledge and turn it into new products that can be manufactured and marketed by the business units, this will contribute positively to the level of perceived external fit.

Searching for the most relevant capabilities, we considered the fact that technology-based prospector companies operate under dynamic market and technology conditions, which force them to continually adapt their competencies and capabilities to changing circumstances. Yamada and Watanabe (2005) point at the importance of R&D adaptability, which is defined as the capability to select the correct R&D themes in response to technological chances and market opportunities, and which is included as 'flexibility' in the present study. Aaker and Mascarenhas (1984) defined flexibility as a firm's ability to adapt to substantial and uncertain changes in the environment. Verdú Jover *et al.* (2005) claim that one of the major challenges of strategic management is to confront environmental change with flexible adaptation, also referred to as agility, in order to reach fit between the firm and its environment. He concluded from a study of 417 European firms in the chemical, electronics and automotive industry that an adequate level of flexibility has a positive effect on the innovative capacity and will lead to higher innovativeness. Because flexibility in the innovation process involves not only the speed by which the existing R&D portfolio can be changed but also how much time it takes before such changes take effect in the form of newly developed products or processes, it was decided to include two dimensions of flexibility: responsiveness (the time it takes to get an innovative idea into the R&D portfolio) and timeliness (the time it takes for an innovative idea to be materialized into project outcomes, once it has entered the R&D portfolio).

A second important capability is communication. Brown and Eisenhardt (1995) categorized the communication web research stream as one of the major streams in product development research, next to innovation as a rational plan and as disciplined problemsolving research streams. The underlying premise of this stream is that communication stimulates performance. The main hypothesis is that the better R&D is connected to the key players outside R&D, the more successful the innovation process will be. Angle (1989) found support for this hypothesis in his study that indicated that innovation effectiveness is related to communication frequency within innovation teams as well as communication frequency outside the teams. It should be noted that most of the studies in the communication web research stream were conducted at project level, while Brown and Eisenhardt (1998) expressed the need for studies at higher levels in the organization. The present study investigates communication at the higher aggregation level of a firm's innovation function: namely the communication within the HQ-R&D-BU triangle by assessing the level of communication in terms of information exchange between R&D staff and their most important BU customers during and after project execution. We expect that the beneficial role of communication, proposed by the communication web stream, will also be evident at this higher aggregation level. We furthermore expect that the hypothesized positive relation between communication and R&D performance will imply a positive relation between communication and external fit. This brings us to the first proposition (P) to be tested in the longitudinal study (L).

P(L)1. A higher level of internal fit, in terms of the perceived adequacy of the R&D competencies and capabilities, will be positively related to a higher level of external fit, in terms of R&D alignment to market and technology needs.

Feedback and strategic alignment

Rosenbloom and Burgelman (1989) state that experience with 'performing' a strategy will have feedback effects on a set of organizational capabilities. The longitudinal study design enables us to observe the effect of feedback loops on the innovation competencies and capabilities within the HQ-R&D-BU triangle by providing the participants with the results of each survey. This works as follows: the structured feedback provided by the longitudinal survey questionnaire at time T_0 will enable R&D and top management to take action with targeted management methods, which will lead to improved and better aligned competencies and capabilities. The subsequent measurement at time T_1 provides information on the effect of these methods and will lead to adjustments or new measures, the effect of which can be measured at time T_2. This brings us to the second proposition to be tested in the longitudinal study.

P(L)2. Structured feedback on R&D competencies and capabilities will help to achieve and maintain internal fit.

Methods to improve alignment

Several methods to improve alignment were introduced during the period under investigation. After the first survey, the management of the corporate R&D center decided to introduce the Balanced R&D Score Card, including financial and non-financial indicators (see Omta and Bras, 2000). Another initiative was the system of market-technology road mapping, which was introduced after the third survey. The most important technical improvement in the period under investigation was the vast increase in the use of computer simulation, which reduced development time by providing the possibility for virtual development and testing.

The most important management measure was taken by corporate headquarters after the second survey, namely a change in the R&D resource allocation structure, from 100% corporate funding to a mixed system of 50% business unit funding and 50 % Technology Board funding. From then on, the business units could use their 50% to find their own R&D projects at the corporate R&D center or, if they preferred, somewhere else. The other 50% was decided upon by the Technology Board, which consists of the CTO, the directors of the business units and the top management of corporate R&D. The Technology Board typically looks at the fundamental projects in the R&D portfolio that invoke higher risk and a longer time horizon, whereas the business units focus more on projects with a shorter time horizon. Based on Buderi (2002), who emphasized the importance of R&D funding, we expect that the change in the decision rights over the way corporate R&D is funded, will effectively balance the long term orientation needed for exploration (via Technology Board funding) and the short term orientation, needed for exploitation (via BU funding). This leads to the third proposition to be tested in the longitudinal study.

P(L)3. An R&D funding structure that effectively balances the exploitation and exploration function will help to achieve and maintain internal fit.

4.4.3 Operationalization of the research variables

To test the conceptual framework (Figure 4.3) empirically, the concepts were split into research variables, which were operationalized by providing operational definitions and indicators to produce measurable items (see Table 4.2). Most research variables were measured by means of items that represent a subjective assessment of the R&D staff and their customers in the business units and headquarters, using five- and seven-point Likert scales, except the variable of internal fit, and the measures taken to improve strategic alignment. Respondents were indicated that they had to regard the intervals between the consecutive values on the scale as equal. Most respondents did, as can be deducted, for instance, from the fact that they sometimes gave 'halves' instead of 'full' figures. In fact, although the scales were intended to be used as interval scales, there is a chance that some respondents may have used them as ordinal scales. This poses no problem for the analysis, however, since we have generally combined individual items into composite variables. In academic practice, it is widely accepted that total scores from individual ordinal items can be treated as interval scales (e.g. Nunnally and Bernstein, 1994).

Strategic alignment was operationalized as a function of internal and external fit. External fit was assessed by the respondents' perceptions of the alignment of the R&D projects to the business environment in terms of technologies and market needs. Internal fit is operationalized as the congruence of the R&D staff's self-perception with the BU customers' perceptions of the level of R&D competencies and R&D capabilities.

To assess the variable R&D competencies, respondents were asked to indicate the relative importance of six R&D objectives ranging from basic research via applied research to

Table 4.2. Operationalization of research variables and measures (question numbers, see Appendix D and Section 4.4.3 for the formula of R&D competencies).

Variables	Operational definitions and indicators	Measures
Strategic alignment		
Internal fit	Congruence of the BU perception and the R&D self-perception on the importance and the achieved level of R&D competencies and capabilities	
External fit	Alignment of corporately and BU-funded R&D projects to the business environment in terms of technologies and market needs	D2 – D5
R&D competencies	R&D lab's achievements weighted for the relative importance of its competencies (basic technology development to applied engineering tasks)	$D1a_{1-6}$, $D1b_{1-6}$
R&D capabilities		
Flexibility: responsiveness	Ease of incorporation of BU requests in the corporate R&D portfolio and time lag to start-up in an R&D project	D6, D7
Flexibility: timeliness	Number of projects delivered before or on the agreed date; the average project cycle time; time lag in answering technical questions	D8-D10
Communication	Amount of contact between corporate R&D staff and the BUs during R&D project execution, after project completion, and after market introduction; clarity of reporting; complaint handling	D11-D15
Methods to improve strategic alignment		
R&D funding structure	Percentage of R&D funding from headquarters versus BU funding	
Virtual development tools	Modeling and computer simulation to replace large prototypes and testing facilities	
Communication improvement measures	Staff exchange and regular R&D-BU and external customer contact to discuss future needs	D16 – D19
Control variables	Business unit, nationality, function (marketing, product and process development) and position (director, manager or engineer), year of response and size of the R&D center	

engineering (1a1 to 1a6, see Appendix D). The respondents were then asked to assess the R&D lab's performance on each of these objectives (1b1 to 1b6). The R&D lab's achievements were subsequently weighted for the relative importance of the different R&D objectives, using the following formula.

$$\{[(1a1-2.5)*(1b1-2.5) + (...) + (1a6-2.5)*(1b6-2.5)]/6 + 4.75\}*7/11$$

In this formula we correct for the fact that without subtracting 2.5 from all scores on the questions 1a and 1b, we would get the most negative result when a technology is perceived of low importance and the achieved level on this technology is perceived as low as well, where this outcome should be more positive than if a respondent indicates that the importance of a certain technology is high, but the achieved level in that particular technology is low, or the other way around. Hereafter, 4.75 was added to get an eleven-point scale and it was multiplied with 7/11 to make the scale comparable to the seven-point Likert scales.

The variable flexibility was measured using two dimensions: responsiveness and timeliness. Responsiveness was captured by the lab's flexibility to incorporate new R&D projects into the corporate R&D portfolio, and the pace of start-up of these projects. Timeliness was measured as the average cycle time of R&D project execution, the number of projects delivered on the promised date and the time lag in answering technical questions.

The research variable communication was assessed as the amount of contact corporate R&D staff had with the business units during R&D project execution, after project completion and after market introduction, and as the clarity of reporting and the level of complaint handling. The methods to improve strategic alignment comprise the change in the R&D funding structure, the use of market-technology road maps, R&D Balanced Score Cards and increased use of integrated virtual development tools, such as modeling and simulation.

In addition, the subjective assessment of the respondents regarding the importance of methods to improve internal fit (regular structured R&D-BU contacts to discuss future needs and staff exchange) and external fit (direct contact of R&D staff with end users) was measured.

The following attributes were used as control variables: (1) whether the respondent came from the corporate R&D center or from the business units or headquarters; and if the respondent came from one of the business units (2) from which business unit; (3) which country; (4) which function (product development, marketing or other); and (5) which position (director, manager or engineer); (6) year of response and (7) the size of the corporate R&D center.

4.4.4 Methods of data collection and data analysis

Data were collected in a multinational supplier of technology-based industrial components for different industries. In 1997, 1998, 2000 and 2002, survey questionnaires were sent to corporate R&D staff and the managers in the BUs and headquarters of the company. To measure the research variables an initial pool of 61 questions was drawn up (approximately 12 items per research variable).

These 61 items were subjected to two stages of data refinement using the same panel of five experts from industry and academia that was also used for the cross-industry study. The first stage focused on reducing the questionnaire by retaining only those items capable of discriminating across respondents, and examining the dimensionality of the scales and establishing the reliability of its components. The second stage was primarily for confirmation and involved re-evaluating the condensed scales' dimensionality and reliability by retesting them.

Some further refinements occurred at this stage. In the six years over which the longitudinal study was conducted, the core questionnaire (30 questions) remained unchanged. The questionnaire used the five-point Likert scale for the R&D competencies, and the seven-point Likert scale for the other variables, ranging from 'strongly disagree' (1) to 'strongly agree' (7), with no verbal labels for the intermediate scale points. Several items were negatively worded to reduce response tendencies by the respondents (Cooper and Emory, 1995). These items were reverse-scored for use in the analyses to ensure that a higher assessment in all cases reflected a more positive judgment of the item at issue.

To investigate possible non-response bias, the answers of late respondents were compared with those of early responders, as Oppenheim (1966) suggested that late respondents resemble more closely non-respondents than early respondents.

The research variables that were investigated in the cross-industry and the longitudinal studies (strategic alignment, R&D competencies, and R&D capabilities, responsiveness, flexibility and communication) are unobservable as such, and must therefore be assessed by means of indicators that can be empirically measured. In this respect, two basic types of variables are relevant: reflective and formative latent variables. Reflective latent variables reflect an underlying construct that can be observed by measuring a number of indicators because the underlying construct causes the observed measures. In the longitudinal study, this was the case for strategic alignment and R&D capabilities. In contrast, a formative latent variable defines the construct. A defined construct is completely determined by the collection of the combined indicators. In the longitudinal study, R&D competencies were defined by the entire range of the technological skills and know-how of the corporate R&D center, ranging from fundamental research to applied engineering. In assessing formative latent variables, the items do not necessarily correlate so the methods of assessing reliability and validity used for reflective variables cannot be used.

The validity tests of both reflective and formative variables start by evaluating the content validity that assesses the degree of correspondence between the items selected to constitute the construct (Hair *et al.*, 1998; Babbie, 2003). As was stated above, an extensive literature search was conducted for both studies to assess the research variables and the relevant items in each variable. In addition, a panel of five experts was asked to assess the content validity of the research variables.

The reliability and validity of the reflective research variables R&D capabilities, strategic alignment, and importance of internal and external communication were analyzed by means of factor analysis to check for the unidimensionality of the scales and the amount of variance extracted for each variable, followed by Cronbach α to check for the internal consistency of the scales. A principal component analysis was conducted to investigate the relationships between the different items, followed by Varimax rotation to reach maximal independency (clusters of items with a high correlation).

The relationships in the conceptual model were tested by means of a linear stepwise regression analysis with listwise deletion for missing values, using the R&D competencies and R&D capabilities as the main independent variables, and the level of external fit as the dependent variable. Also other linear regression analyses methods -enter, forward and backward regression- were tested, but this did not chance the conclusions. Whether interaction occurred among the different research variables was also investigated (Baron and Kenny, 1986). To investigate whether respondents who participated more than once might tend to give a more positive judgment, based on feedback from the earlier survey, the analyses were checked using the independent sample of the total database (n=474), meaning that every respondent was entered only once. These analyses were carried out using the first time, and the last time that a frequent responder participated. None of these analyses provided different results than using the whole sample. For this reason it was decided to use the whole study sample of 696 respondents for the analyses, and the results of this analysis years (1997, 1998, 2000 and 2002) were introduced as dummies. The number of respondents from the headquarters was too small (3 to 7 respondents per survey) for separate analysis. They were therefore added to the BU respondents. To analyze the gaps between the assessments given by respondents from corporate headquarters, the BUs and the self-assessments given by R&D staff, two-tailed t-tests were used. We chose for two-tailed t-tests because no propositions were developed concerning the direction of expected relationships. Non-parametric analyses of group means, using the Mann Whitney Test, did not alter the conclusions.[1]

[1] Stevens (1946, 1951) proposed that measurements fall into four major classes: nominal, ordinal, interval and rational and that these levels allow progressively more sophisticated quantitative procedures to be performed on the measurements. Nunnally and Bernstein (1994) however argue that the results of summing (ordinal) item responses are usually indistinguishable from using more formal methods, especially when dealing with larger (>100 items) datasets, and that Stevens' position can easily become too narrow and counterproductive. Stevens' representational theory is also being disputed on more fundamental grounds. For an in-depth discussion, the reader is referred to Gaito (1980).

4.5 Concluding remarks

In this chapter, the general conceptual framework and the overall research design have been presented. We elaborated on how the two approaches to investigate strategic alignment in innovation, identified in Chapter 3, led to the design of two consecutive and complementary empirical studies, namely the cross-industry study and the longitudinal study. In the cross-industry study, the product generation life cycle was proposed as the indicator to assess the combined effect of the factors of market dynamism and technology complexity, as derived from the industrial organization theory. The propositions regarding the expected differences in innovation strategy and strategic alignment of innovation to business between companies with relatively long product generation life cycles (LLCIs) and relatively short product generation life cycles (SLCIs) were also discussed. The essential element of the longitudinal study, designed to investigate the factors derived from the competence perspective, is the comparison of perceptions of R&D staff, and those of the managers in headquarters and the business units regarding the quality of the R&D competencies and capabilities to fulfill their company's long-term and short-term objectives effectively and efficiently. The feedback provided by the longitudinal survey questionnaire could be used to improve alignment, defined as the level of congruence between the self-perception of R&D staff on the one hand, and the perception of managers in the BUs and corporate headquarters on the other. Finally, the research variables, and the methods of data collection and analysis were discussed for both studies.

5. Cross-industry study results[2]

This chapter reports the results of the cross-industry study that was set up to answer the first research question.

RQ1. What is the effect of the industry 'clockspeed' on the strategic alignment of innovation to business?

Section 5.1 discusses the data collection and provides a baseline description of the participating companies. The R&D performance of the participating companies is presented in Section 5.2, using the revenue contribution of new products and the number of patents per R&D investment as indicators. The revenue contribution of new products is also used to support the division between SLCIs and LLCIs. In Section 5.3 the differences in strategic alignment of innovation to business between LLCIs and SLCIs are reported. Section 5.4 elaborates on the differences found between LLCIs and SLCIs in the level of exploration of R&D strategy, and R&D portfolio planning. Section 5.5 discusses the planning of the internal and external R&D competencies (open innovation) and Section 5.6 elaborates on the R&D capabilities, timeliness, responsiveness, and internal and external communication. This section also provides in-depth descriptions of the management practices used in the R&D centers under investigation because of their importance for practitioners in innovation management. Section 5.7 focuses on the R&D process, and governance structures are discussed, such as R&D project organization and the systems of collegial network steering in place in some of the participating companies. In Section 5.8 a number of methods to improve strategic alignment and their use in the participating companies are discussed. Finally, Section 5.9 provides the reader with some concluding remarks.

5.1 Data collection and baseline description of the cross-industry companies

There turned out to be widespread interest in comparing companies' own R&D management practices with those of leading companies in other industries, and ten of the fifteen companies approached agreed to participate in the cross-industry study, for instance, Airbus, Erickson, Exxon and Philips. In two cases we contacted two companies in the same industry. This turned out to be unnecessary; and in both cases we had to halt our contacts with the second company since the other had already indicated being interested in participation. Of these ten companies, eight are headquartered in the EU and two in the US. Furthermore, four come from Short Life Cycle Industries, defined as industries characterized by PGLCs of (much) less than six years, and six companies are from Long Life Cycle Industries (PGLCs > 6 years). In this we follow Williams (1998) who classified companies into fast-cycle (PGLC < 2 years), standard-cycle (PGLC < 6 years) and slow-cycle industries (PGLC > 6 years). In his classification

[2] In 2007 a paper was published reporting on the cross-industry study, Fortuin and Omta (2007a).

he included non technology-based industries, such as fast food and furniture, which were generally standard-cycle. Because the present study focuses on technology-based industries it was decided to combine the fast-cycle and the standard-cycle into one group, that of the SLCIs.

The SLCI companies are from the following industries: copier and printing technology, domestic appliances, electronics, and mobile phones. The LLCI companies come from the aircraft, aerospace, industrial equipment and components, energy and pharmaceutical industries. In total 16 structured interviews were held with CTOs, Directors of Corporate R&D centers and technology directors, and 30 completed questionnaires were returned by the department heads and/or program managers.

Table 5.1 gives indicative figures from the participating companies. Four of the companies are big, with sales volumes of over US$ 50 bn, and employee numbers rising to nearly 370,000. At the time of investigation, the participating companies were growing rapidly at an annual rate of about 10%. One of the companies was growing much faster, showing a dazzling 42% growth in the preceding three years. The operating profit margins differed considerably from 2.5% to 17%, However, to assure a valid interpretation of this figure, the differences between industries must be taken into account.

Figurc 5.1 and 5.2 compare the cross-industry companies with other technology-based companies. Figure 5.1 shows the percentage invested in R&D. At the low end, we find industries with a high sales volume compared to the investments needed, such as in the oil industry. At the high end there are technology-intensive industries, such as the aircraft and pharmaceutical industries. Figure 5.2 compares the operating profits of the companies under study with industry averages. For most of the participating companics, total R&D expenditures are below US$ 300 m, only three companies spent much more, over US$ 500 m. R&D expenditures as a percentage of sales range from 0.4 to 16% (see also Table 5.1). The number of R&D staff refers to the R&D center that was examined and not to the total number

Table 5.1. Baseline description of the participating companies, range and average (between parentheses).

Sales volume	1-20 US$B (6 comp.), > 50 US$B (4 comp.)
Number of employees	19,000 - 370,000 (69,000)
Annual growth rate	7 - 42% (9.8%)
Operating profit margin	2.5 - 17% (8.9%)
R&D expenditures	50 – 300 US$M (7 comp.), > 500 US$M (3 comp.)
R&D as % of sales	0.4 -16% (5%)
R&D staff in participating lab	110 to 1800 (700)

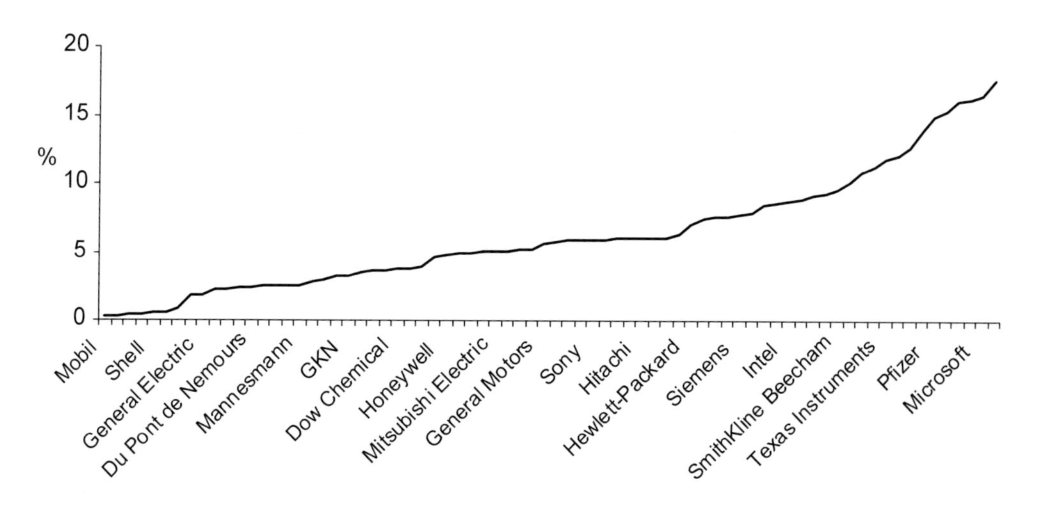

Figure 5.1. R&D as a percentage of worldwide sales in 77 technology-based companies, averaged over 1996 and 1997.
Source: Annual reports, Business Week 02-02-1998 and 10-07-1997, Davis (1997), Nordicum, The Scandinavian Business Review, 4-1997, The 1996 and the 1997 UK R&D Scoreboard, Department of Trade and Industry, Company Reporting Ltd., Edinburgh.

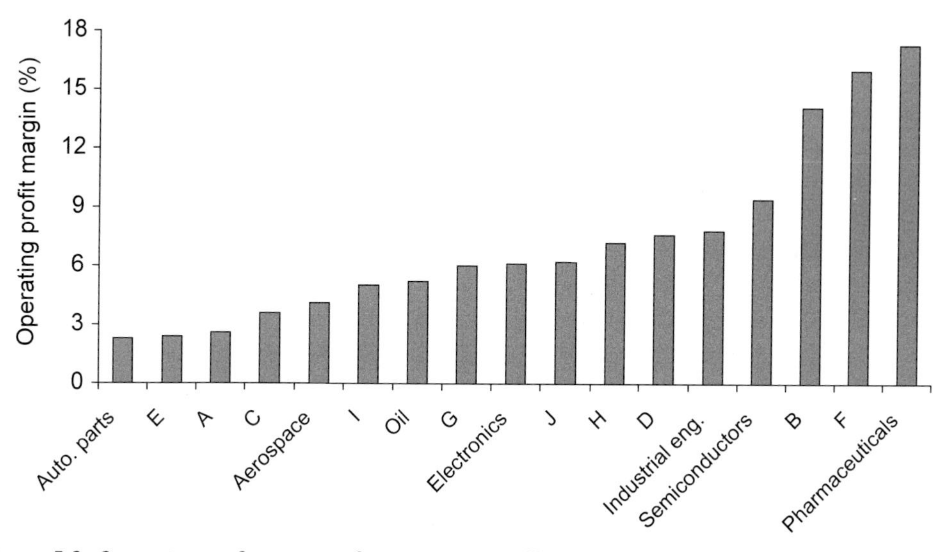

Figure 5.2. Operating profit margin of companies in different US industries and including the cross-industry companies (A to J).
Source: Annual reports, Business Week 02-02-1998 and 10-07-1997, Davis (1997), Nordicum, The Scandinavian Business Review, 4-1997, The 1996 and the 1997 UK R&D Scoreboard, Department of Trade and Industry, Company Reporting Ltd., Edinburgh.

of R&D staff employed by the company because we concentrated on one R&D center only, namely the corporate R&D center or on the R&D center that concentrated on one of the major operating areas of a diversified company.

5.2 R&D Performance

Two indicators were used to assess the innovative performance of the companies that participated in the cross-industry study: the revenue contribution of new products was used as an indicator of the commercial output of R&D; and the number of patents per R&D investment was used as an indicator of the research phase of the R&D process.

5.2.1 Revenue contribution of new products

Figure 5.3 shows one of the most important indicators of innovative output, namely the revenue contribution of new products (the contribution of sales revenue derived from products introduced on to the market in the last three years by the different BUs of the participating companies against the average PGLC of the companies). The ten companies indicated the average revenue contribution of new products for 17 strategic business units (one to three BUs per company). Although within a company the PGLC differs slightly per BU, for clarity of presentation of each company the average PGLC was calculated, so all BUs in one company show the same PGLC length.

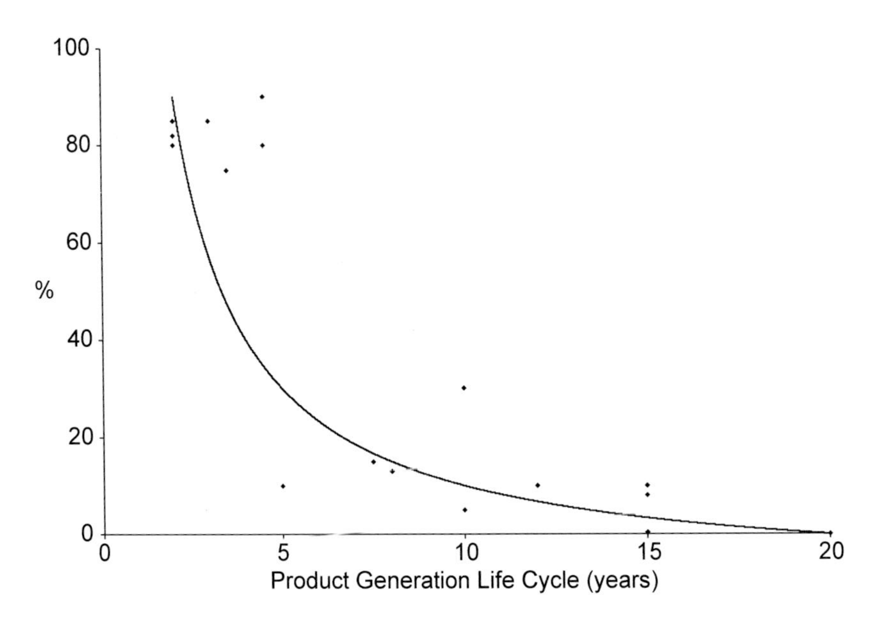

Figure 5.3. Revenue contribution of new products introduced on the market by the different BUs against the average product generation life cycle (PGLC) per company (years).

Theoretically, the revenue contribution curve would be expected to decrease hyperbolically from 100% towards the X axis. In its simplest form, if revenue contribution is assumed to be constant over the whole PGLC length, it can be modeled by the following formula:

Revenue contribution (%) = 100% x (3 years) / PGLC (years)

This simple formula fits the data in Figure 5.3 of the SLCIs quite well, but as the PGLC increases it shows a tendency to overestimate revenue contributions. However, the model can be refined by adding a parameter for the delay caused by the fact that sales volumes need time to reach their full potential after market introduction. We assume that this delay increases with increasing PGLC length. The curve in Figure 5.3 assumes a delay of (1 year + 0.1xPGLC). Described by the following formula:

Revenue contribution (%) = 100% x {3 years - (1 year + 0.1 x PGLC)} / PGLC

Figure 5.3 shows that there is a clear gap between SLCIs and LLCIs with PGLC length of six years. Companies in industries with product generation life cycles of up to five years reported revenue contributions of new products of 75% or more (with one exception). Companies with longer life cycles between the different product generations reported lower revenue contributions, but not zero. This was even the case for companies in industries where the life cycles are very long, such as in the aerospace and the oil industries. The reason for this is that the life cycle of, for example, an airplane might be long but sub-systems within the airplane, such as the electronics system, are replaced at much shorter time intervals.

The highest revenue contribution scores for each PGLC in Figure 5.3 can be used as benchmarks, and companies and BUs that score lower might find it advisable to accelerate the pace of new product introduction. Indeed, in the structured interviews one of the CTOs complained about the under-achievement of one of the BUs in this respect compared to competitors. When analyzing the data later, this BU did come out below the benchmark. Such underachievement may not be caused by the BU and the R&D center alone; the corporate headquarters may also play a role. As we will see in Table 5.3, not many Corporate Boards, especially in LLCIs, stimulate their BUs to generate part of their sales from new products and processes.

5.2.2 Number of patents

An important indicator for the outcome of the research phase is the number of patents per R&D investment. Interestingly, in clear contrast to Proposition P(C)2 (see Section 4.3.4 and Table 4.1), the number of patents is about six times higher in SLCIs than in LLCIs (3.3-9.2 patents per US$ 10 m in SLCIs versus 0.5-1.5 per US$ 10 m in LLCIs). This unexpectedly high number of patents per R&D investment may reflect the essential position that industrial property rights have gradually attained in SLCIs. In the structured interviews, the CTO of one of the SLCI companies explained this emphasis on patenting as follows.

In order to prevent reversed engineering, the legal strategy should be directed to make it extremely difficult to copy your products and processes. Furthermore, the increased emphasis on R&D efficiency necessitates that vital parts of modules have to be reused in the next product generation(s), so we need longer patent protection.

Knowledge ownership basically refers to the maintenance of the knowledge base of the firm, i.e. patent strategy and the eventual need for defensive research. The structural interviews indicated that in all the companies the importance of patent strategy has increased in the last decade. The income, either gained or paid in licensing, and the penalties stemming from litigation have convinced business that intellectual property rights are a major concern. The cross-industry companies submit on average one patent (0.92) per ten R&D staff members annually.

Most participating companies use a rigorous internal screening process to determine the patentable ideas of their R&D staff, based on the technical literature and peer review of technical and legal experts. Two-thirds of the participating companies have patent committees that meet regularly to prioritize invention disclosures. They concentrate on the following questions.

- Is this idea new and novel, and is there a potential for commercial utility?
- Does it work on R&D scale and is it likely to work in practice?
- Do we want to patent it or is it better not to reveal it?
- If so, is it necessary to patent it worldwide at increased cost or is it better to patent it in specific regions only?

Typically 35% to 55% of ideas survive the internal screening process, whereas 90% to 98% of the filed ideas eventually lead to patents being granted. This is many more than reported by Stevens and Burley (1997), who estimated that typically 1 in 10 ideas survives the internal screening process. They further conclude that patent fees comprise about 1% to 2% of the R&D budget of major US companies. About 8% of the issued patents turn out to create some value and 1% creates major commercial value for the firm.

A number of the participating companies use some kind of patent tracking indicators, such as the number of new patent applications per year; the percentage of patent applications filed vs. abandoned, and/or royalties received vs. patenting costs. In some of the companies a great deal of attention is paid to patent cycle-time reduction. Each step in the patent filing process is analyzed to determine whether it is necessary and whether steps can be done in parallel rather than in sequence. For this purpose, special time sheets are used, e.g. time from internal invention disclosure to action by the attorney, time to patent submission, and time to filing.

5.3 Strategic alignment

5.3.1 External fit

Table 5.2 shows that, in accordance with Proposition P(C)1 (see Section 4.3.4 and Table 4.1), all aspects of external fit are assessed to be significantly higher in SLCIs than in LLCIs. Companies in SLCIs, which encounter high market dynamism, clearly put more emphasis on monitoring market and technology trends than companies in the more stable LLCIs. Market-technology road maps are used more intensively in SLCIs.

Table 5.2. Self-assessment of the level of external fit, seven-point Likert scales, mean and (standard deviation, s.d.).

	SLCIs	LLCIs
Market and technology trends are important strategy inputs	6.7 (0.5)	5.6 (1.1)**
Core technologies are well defined	6.7 (0.5)	5.4 (0.5)***
Portfolio is based on a strategic technology/market vision	6.4 (0.5)	4.8 (1.1)**
Market-technology road maps are updated regularly	6.2 (0.8)	4.7 (1.5)**

** $p < 0.05$;
*** $p < 0.01$

5.3.2 Internal fit

To assess the level of internal fit (see Figure 4.1), we asked the R&D managers for their opinions about the alignment of the innovation strategy with business interests, and about the judgment of the BU managers on this issue.

We used the R&D managers' estimations about strategic alignment of innovation to business as a proxy for the actual business unit opinion. Table 5.3 shows that, in accordance with Proposition P(C)1, R&D management in SLCIs is more positive about their alignment to business strategy than in LLCIs.

This is especially the case for participation in the establishment of business plans and the other way around, although to a lesser extent, the involvement of BUs in R&D planning is more positively assessed in SLCIs. However, differences concerning the monitoring of key project parameters and the aligning of R&D objectives with business plans are small.

Table 5.3. Self-assessment of the internal fit between innovation and business strategy, the R&D viewpoint, seven-point Likert scales, mean and (s.d.).

	SLCIs	LLCIs
Close linkage of R&D and business strategies	6.2 (0.6)	5.7 (0.9)
R&D project evaluation aligns with business plans	5.6 (1.1)	5.5 (1.5)
R&D project parameters are monitored with business	6.2 (0.5)	6.3 (0.6)
R&D objectives are set in line with business plans	6.0 (0.4)	6.4 (0.6)
R&D participates in the establishment of business plans	5.7 (0.6)	4.4 (2.0)*
Cross-functional participation in R&D planning	5.4 (1.5)	5.0 (1.3)

*$p < 0.1$

Table 5.4 shows a less positive picture. Here the R&D department heads and R&D program managers were asked to assess the way that they thought the work of corporate R&D center was appreciated in the BUs. The results show that especially the LLCI respondents thought that there were clear frictions in the relationship between R&D and the business units from the BU point of view. Table 5.4 further shows that corporate support to reinforce the R&D interest of BUs, for instance, by 'forcing' them to generate part of their sales from new products and processes, and stimulating them to start high-priority joint projects with R&D, were absent in most of the participating companies.

Table 5.4. Self-assessment of the internal fit between innovation and business strategy, the BU viewpoint, seven-point Likert scales, mean and (s.d.).

	SLCIs	LLCIs
Assessment of R&D's contribution to corporate business	5.2 (1.1)	3.4 (1.5)**
BUs forced to generate sales from new products/processes	5.1 (1.8)	3.2 (1.4)*
BUs select high-priority joint effort R&D projects	3.9 (1.5)	3.9 (1.5)

*$p < 0.1$;
**$p < 0.05$

5.4 R&D strategy

5.4.1 Degree of exploration

Table 5.5 presents the results of the quantitative Questionnaire I (see Appendix B, and further referred to here as quantitative data) on the companies' degree of exploration. It shows that the R&D centers in LLCIs tend to have a more exploratory, long-term technology focus, as proposed in Table 4.1. These companies typically spend 40% or more of their total R&D budget on research and technology development, including all basic and applied research activities directed towards expanding the company's technological knowledge base. These research activities are often directed towards the development of future projects but occur prior to specific product and/or process development activities.

In contrast, R&D centers in SLCIs typically spend 80% or more of their total R&D budget on product and process development, and engineering and testing. The difference in the degree of exploration is also reflected in the educational level of the R&D staff in the participating R&D centers, approximated by the number of R&D staff with a PhD. In LLCIs 18% to even 30% of the R&D staff have a PhD, whereas in SLCIs this percentage is much lower, usually between 3% and 6%. In the structured interviews the zero-defect situation that is so important in LLCI companies was also stressed by the CTO of a satellite communication company, who stated: *the launch of a satellite is like shooting a whole plant into the sky. Because it goes far into the galaxy, we have to take care for 15 to 20 years of predictable reliability.*

Table 5.6 shows that the R&D managers in SLCIs as well as in LLCIs indicate that their R&D labs are more directed towards the exploitation of existing knowledge by focusing on incremental improvements of products and processes than on gaining new knowledge needed for 'radical' breakthroughs. We find a somewhat more radical orientation in LLCIs but it is far from significant. R&D managers apparently in LLCIs have the feeling that there is not enough room for rethinking the fundamentals of their industries. Table 5.6 shows another unexpected finding, especially for SLCIs, namely that the pace of conducting R&D projects is not emphasized over budget. Table 5.6 further shows that senior management in LLCIs devotes more, although not significantly more, attention to concept specification and planning

Table 5.5. Degree of exploration in short and long life cycle industries (SLCIs and in LLCIs), quantitative company data, Questionnaire I.

	SLCIs	LLCIs
Basic research as % of total R&D expenditures	13-20%	40-60%
PhDs as % of total R&D staff	3-6%	18-30%

Table 5.6. Degree of exploration. Qualitative self-assessment by R&D department heads and R&D program managers, Questionnaire II, seven-point Likert scales, mean and (s.d.).

	SLCIs	LLCIs
Incremental advances are more important than breakthroughs	4.1 (2.0)	4.6 (0.9)
Funding for fundamental research is relatively easy to get	4.7 (0.3)	2.9 (1.6)**
Speed is emphasized over budget	4.6 (0.8)	4.3 (2.0)
Senior management devotes attention to early development [a]	4.4 (1.5)	5.0 (1.1)

**$p < 0.05$
[a]rotated scale

than senior management in SLCIs. However, if we concentrate on value engineering, the relationships are clearly in line with our expectations. Value engineering refers to the creation of the highest customer value for the lowest cost by implementing improvements in product or process design.

Table 5.7 shows in accordance with Proposition P(C)2 that the attention paid to value engineering in the participating companies is significantly higher in SLCIs. The participating companies indicate having achieved about 15% (5% to 30%) savings in their present products and processes by changes adopted in the last three years.

5.4.2 R&D portfolio planning

The strategic importance of R&D in the participating companies is indicated by the fact that the main responsibility for the R&D portfolio is high up in the organization. In a number of the participating companies, the Board of Directors is in some way involved in the prioritization process. In the structured interviews with the CTOs, the directors of the R&D centers and

Table 5.7. Self-assessment of attention paid to value engineering, seven-point Likert scales, mean and (s.d.).

	SLCIs	LLCIs
Cost drivers and capital constraints taken into account	6.2 (0.8)	5.3 (0.9)*
Opportunities for cost savings consistently considered	6.7 (0.5)	5.8 (1.4)
Attention for product/process robustness and cost-effectiveness	6.2 (0.3)	4.8 (1.1)**

*$p < 0.1$;
**p , 0.05

the technology directors (see Appendix A, further referred to here as structured interviews) it became apparent that all participating companies consider the following factors in their portfolio screening process.

- Compatibility to business goals and the desired R&D balance between long-term and short-term objectives.
- The availability of in-house technical skills and facilities, and the probability of technical success.
- Timing of R&D and market development relative to competition. Potential market size and stability of the potential market to economic changes.
- Investment level and potential rate of return of R&D projects.
- Knowledge protection and the need for further defensive research.

In most participating companies the R&D project portfolio is divided into two major groups of projects:

- exploratory projects with an outlook of five to ten years to fulfill the technology needs of the business; and
- business-connected projects directed toward the needs of the business units.

The selection was then based on the following criteria:

- What technologies will the business need to compete in the coming years?
- Do we have to develop them ourselves, should we purchase them, or should we develop them with business partners or knowledge institutions?

After discussions with senior management at company headquarters, the business units and the R&D Center, these questions were discussed in a Research Guidance Conference.

Most participating companies reported that they were conducting too many R&D projects and too many of these involved too great a risk. They indicated they had critically reviewed their R&D portfolios and had not only trimmed the total number of their projects but had also balanced the risk in their project portfolios. They indicated that they spent not more than 10% to 15% of their total R&D budget on ('new-to-the-world' or 'new-to-the-firm') pioneering projects and 50% to 80% on lower-risk major projects. Pioneering projects are R&D projects in which the product and/or process technologies are developed and implemented for the first time. These projects establish the starting point for new manufacturing processes. The rest of the R&D budget is spent on minor projects and technical consulting tasks. However, one LLCI lab reports spending over 80% (!) of its total R&D budget on 'new-to-the-world' projects. If we compare the percentage of R&D spent on pioneering projects with the figure for high-risk projects found in the 1995 PDMA survey (Griffin, 1997), we see that this percentage is lower. In this survey the companies reported spending on average 25% to 30% of their total R&D budget on projects that were either 'new-to-the-world' or 'new-to-the-firm'. Taking possible differences in definition and interpretation into account, we may conclude that the participating companies are at least as good, and possibly better, at balancing risk in their project portfolios. On the other hand, the participating companies might have gone too far, and while minimizing risk in their project portfolio have also diminished their chance of finding real breakthroughs. A risk that was clearly mentioned in a number of the structured interviews. Therefore, the one company in our survey that is proceeding in the opposite direction might turn out to be the winner in terms of innovation in the long run.

5.5 R&D competencies

5.5.1 In-house R&D competencies

Knowledge management constitutes an important aspect of innovation strategy implementation. Deciding on an appropriate form of knowledge management includes making a planning for the necessary level of R&D competencies. Since R&D in a company has a bridge function between the world of science and technology and the business world, R&D has to monitor new and potentially fruitful ideas and technologies, transform them into knowledge useful for the company, capture them in knowledge databases and make them available to the BUs. To fulfill this role effectively R&D has to define its core competencies, and needs systems to bridge the gap with the business world. Examples of core competencies mentioned in the structured interviews are the ability to work with real-time systems; the ability to design large and complicated systems with good functionality; and the breakdown of products into modules with the right interfaces.

Table 5.8 shows the level of in-house R&D competency planning as perceived by the respondents. It shows that, in accordance with Proposition P(C)3, SLCIs put more emphasis on competency planning. The cross-functionality of competency monitoring in particular is assessed significantly higher in SLCIs than in LLCIs. In the structured interviews some respondents indicated that they found competency listing to be especially useful for the recognition of gaps, i.e. the core competencies that should be acquired to achieve market success. This makes the competence list a vision pulling the company forward, rather than a snapshot that keeps it anchored in the past (Griffin, 1997).

Table 5.8. Qualitative self-assessment of the level of the in-house R&D competencies, seven-point Likert scales, mean and (s.d.).

	SLCIs	LLCIs
Existence of cross-functional competencies monitoring system	6.0 (1.1)	5.1 (0.8)*
Efficient exploitation of R&D competencies	5.1 (0.8)	4.4 (1.1)
Regular monitoring of possible competency gaps	5.9 (0.6)	5.3 (0.9)

*$p < 0.1$

5.5.2 External R&D competencies, open innovation

Companies do not only have to organize their internal R&D competencies. Increasingly they are working together in networks of ever-increasing complexity. It is the emerging challenge of R&D management to balance internal and external R&D competencies. Decisions have to be made regarding which competencies are to be kept in-house and which can be outsourced, and how much is done in collaboration with outside partners.

Table 5.9 shows the level of open innovation as perceived by the respondents. It shows that, in contrast to Proposition P(C)3, not much difference was found between SLCIs and LLCIs regarding the level of cooperation with suppliers and customers. But in accordance with Proposition P(C)3, LLCIs make significantly more use of specialized R&D contractors, such as research firms and knowledge institutions. In the following sub-sections we elaborate on how open innovation is implemented in the participating companies, based on the quantitative data provided by the companies in Questionnaire I, and information gathered during the structured interviews.

Table 5.9. Qualitative self-assessment of the level of open innovation, seven-point Likert scales, mean and (s.d.)

	SLCIs	LLCIs
External relations are considered for the execution of projects	5.6 (1.0)	5.7 (1.1)
Importance of developing external R&D linkages understood	4.8 (0.6)	5.3 (1.2)
Regular use of specialized R&D contractors	4.4 (1.3)	5.6 (1.5)*
Early involvement of suppliers	4.8 (1.0)	5.0 (1.3)

*$p < 0.1$

Strategic alliances

The structured interviews indicated that most of the participating companies were involved in a number of strategic alliances to share development costs, accelerate product and process development and maximize commercialization opportunities. It was generally agreed that the information obtained through these alliances had been migrating down through their organizations, leading to a stronger technology knowledge base. In the pharmaceutical industry about a third of development work is carried out in alliance with others, mostly biotech companies. From the interviews it became apparent that the motives for starting a partnership were the following:

- to accelerate time-to-market;

- to improve the cost-effectiveness and quality of R&D;
- to develop stronger technology competencies;
- to broaden the scope of technology reach and/or geographic coverage; and
- to get up-front information about potential acquisitions.

These external relationships might be a reasonable response to business pressures, but at the same time they may create new long-term dependencies and vulnerabilities, as companies are becoming increasingly dependent on outside sources for their technological advances. For instance, if industry is going to entrust critical parts of its research to outsiders, there must be confidence about timing, cost-effectiveness and security and relevance of results. In the interviews it was stated that if a company is entering into a collaborative venture, it always wants to keep the competitive edge. Therefore, any really competition-sensitive research is normally done in-house. The respondents indicated that a good contractual arrangement is also very important. It should at least codify:

- the financial and personal responsibilities of the partners;
- the division of any possible gains among the partners;
- penalty clauses to discourage opportunistic behavior;
- the method of knowledge protection, including patent and trade secret rights, and confidentiality agreements; and
- criteria for measuring and monitoring progress, so that deviations can be identified and potential problems overcome. This includes milestones and project deadlines, responsibilities and accountability of the project team and the founding of a control committee. For example, a contract between a biotechnology and a pharmaceutical company included a list of the principal scientists who would be responsible, a detailed schedule of at least weekly telephone conferences, and provision for quarterly joint meetings.

Co-development and cooperation with knowledge institutions

Good supplier handling is generally seen as a key to improving time-to-market. In most participating companies suppliers were gradually evolving from simple producers to co-developers with joint R&D responsibility. The supplier usually had to take responsibility for jointly developed parts. There was no up-front guarantee that the supplier would actually deliver the parts. In companies that were involved in co-development projects, the need for a joint ICT platform was expressed.

Some of the interviewees complained about cooperation with the large Research and Technology Institutes (RTIs) in the area of more fundamental problems. An unexpected observation was that the large RTIs are not always real centers for technology development. In contrast with their mission statements, few carry out leading-edge industrial research. The most successful RTIs focus on highly specialized, but not-so-advanced, engineering and experimental development work carried out in long-term cooperation with industry.

Cooperation in (supra-) government-funded R&D projects

In the companies studied, government- and supra-government-funded (i.e. EU) basic research ranges from 0% to 20% of the research budgets. In general, government-funded development projects are clearly fewer in number, because most government funding is directed towards pre-competitive research. Some of the companies that received (part of) their revenue from the military sector got a considerable percentage of their R&D budget directly from government. Opinions about the value of government funding are not as positive as might have been expected. A number of participants in the structured interviews indicated that government funding in most cases was not directed towards business needs. Another problem with governmental funding is that the benefits must be shared with other companies. For a US company that previously used to do government-initiated research, this was one of the reasons for opting out. European companies in the present study reported that the real value of EU programs was to establish contact with (foreign) companies and knowledge institutions. In particular the staff exchange that results from such transnational cooperation is seen as an important tool in increasing the technology knowledge base of a firm.

5.6 R&D capabilities

5.6.1 Timeliness

As was discussed in Section 2.6.2, R&D capabilities include the managerial and technical systems that exploit those reservoirs in delivering value to customers. Examples of capabilities mentioned in the structured interviews were the ability to manage large scale projects; supplier integration; pipeline management; and information management to record the worldwide technology knowledge base.

The respondents emphasized the importance of identifying 'business' capabilities. For instance, one of the participating R&D centers identified systems marketing, i.e. the ability to sell complex products to technically sophisticated and organizationally complex customers, as one of their core capabilities. This capability enabled them as an R&D lab to enter new markets, for instance in developing new automotive products.

An important measure of the timeliness of R&D and an important improvement target, especially in SLCIs, is project cycle time, the time spent on an R&D project from the start of the conceptualization phase to the launch to the (internal or external) customer. The average cycle time of typical major R&D projects in the cross-industry study is one to two years in SLCIs and two to four years in LLCIs. All participating companies reported having problems with the timeliness of their R&D projects. They indicated that from the moment of release to manufacturing only a third to a half of R&D projects were still on schedule. Figure 5.4 shows that the cycle time reduction realized in the preceding three years in LLCI companies was considerably higher than in SLCI companies. This was an unexpected finding, and in clear contrast to Proposition P(C)4, seen in the light of the vital importance of speed in SLCIs. That these differences were not mere coincidences is indicated by the fact that the cycle time reductions found in the present study were in line with those found in a much larger group of 200 business units in SLCIs (McGrath, 1995). Some of the R&D centers in SLCIs reported the same percentages of cycle time reduction as the average of their respective industries. Their data are therefore not shown separately in Figure 5.4.

Perhaps this unexpected result shows that the R&D centers in LLCIs are catching up with those in SLCIs that have already achieved major improvements in cycle time reduction in the preceding period. Since it takes considerable effort to reduce R&D cycle times further when the first major steps have already been taken. Further investments that the SLCI companies apparently do not want to make, since they mention that speed is not emphasized over budget (see Table 5.6). That timeliness is thought to be under control in the automotive industry will be shown in Figure 5.6.

In the structured interviews, a CTO of one of the SLCI companies indicated that his company has set double digit cycle time reduction targets for the years ahead. In the interview it was claimed that the huge investments needed were accepted in order to gain a further market share with new and innovative products in the extremely turbulent market. The interviewee pointed to an analogy with the Olympic Games, where it always proves to be possible to push the records to heights that were inconceivable only a few years before. In the interviews it was indicated that cycle time could be further reduced, by:

- improved international communication to avoid overlap between R&D centers;

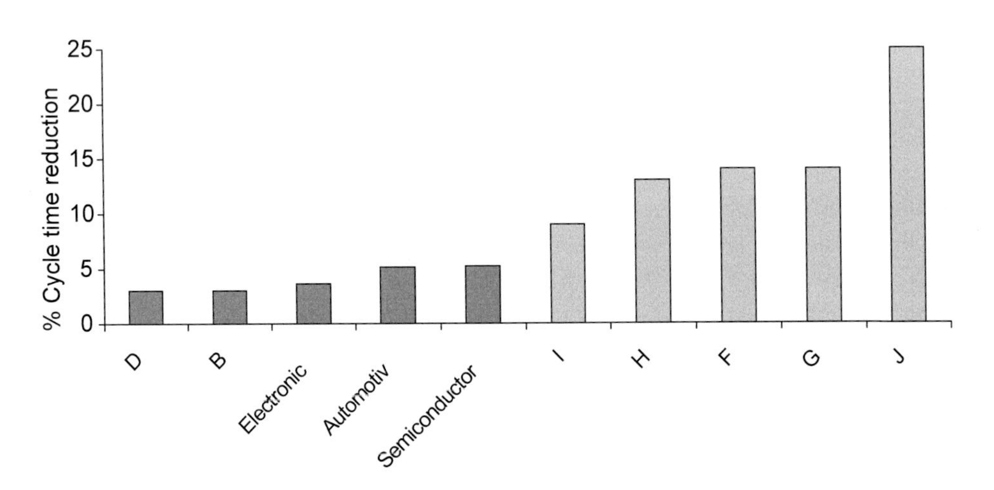

Figure 5.4. Percentage of cycle time reduction in the preceding three years realized by cross-industry companies compared to industrial averages (dark = SLCIs, light = LLCIs).
Sources: the present study and McGrath (1995).

- improved transparency in R&D planning;
- stricter time schedules and milestones to reduce avoidable in-house slack;
- improved use of virtual development tools and further streamlining of ICT platforms; and
- co-development and outsourcing.

5.6.2 Responsiveness

Another important factor is responsiveness, indicated by the time lag between a request for an important R&D project by one of the BUs and its actual start at the R&D center, and the time it takes to reallocate a substantial part (20%) of the R&D resources to a new research area.

It was expected that the R&D centers in SLCIs would be more responsive (see Table 4.1), because of the higher market dynamism they are confronted with, but no actual difference was found in the answers between labs in SLCIs and LLCIs. All participating companies indicated in Questionnaire I that they reacted quickly to BU requests. Typically, a major project can start within nine months, a minor project within three months and an extended task within two weeks of a BU requesting it. Two of the smaller R&D centers reported that it would take only one to three months to move to a new research area. In the larger labs such a shift in resources takes more time but only one of the R&D centers thought that such an operation would take more than a year. A major shift in research orientation had happened in one of the LLCI companies: a shift in R&D capabilities was being made from plasma physics into wireless communication. The whole shift in orientation was completed in less than two years. To be able to move to a new research area quickly, it is important that the procedures for the appointment of new R&D personnel and the purchase of advanced R&D equipment do not take too much time. All companies reported being able to finish both procedures within six weeks.

5.6.3 Internal communication

Table 5.10 shows that, in accordance with Proposition P(C)4, the level of internal communication with the divisions, manufacturing and marketing is consistently assessed as higher in SLCIs than in LLCIs. In the case of communication with divisions and manufacturing, these differences are significant. It further shows that the assumed negative assessment by the BUs of the contribution of their R&D centers, as was shown in the section on strategic alignment (Section 5.3), cannot be attributed to a lack of communication structures as was mentioned in the self-assessment by the R&D department heads and R&D program managers. They were satisfied about the communication systems that were in place, especially with the divisions (a 6.9 on a seven-point scale in SSCIs and a 6.4 in LLCIs). The communication with manufacturing and the BUs was assessed somewhat lower, especially in LLCI labs.

Table 5.10. Self-assessment of the level of cross-functional communication, seven-point Likert scales, mean and (s.d.)

	SLCIs	LLCIs
Project progress communicated regularly to the divisions	6.9 (0.3)	6.4 (0.8)*
Effective communication structures with the divisions	5.9 (1.1)	6.4 (0.5)
Regular interaction with internal (BU) customers	5.3 (0.5)	4.2 (1.7)
Good communication with manufacturing	5.5 (0.9)	4.6 (0.5)*
Good communication with marketing	4.8 (1.7)	4.0 (1.4)
Market information regularly passed on by marketing	4.9 (1.3)	4.3 (1.3)
Competitor information regularly passed on by marketing	5.0 (1.4)	4.4 (1.5)
Meetings with marketing held regularly	5.0 (0.9)	5.0 (1.4)

* $p < 0.1$

Table 5.10 also shows that in many of the R&D centers there was a feeling that cross-functional communication with marketing should improve. More information - both about markets and competition - should be passed on by marketing and the frequency of meetings with marketing should be increased. Many CTOs indicated in the structured interviews that a great deal of effort had been put into closing the R&D-marketing gap. A CTO in one of the LLCI companies indicated that the communication between R&D and marketing had been greatly improved by special programs directed towards building up knowledge about each others expertise and creating a situation of mutual trust. R&D and marketing have now unlimited access to each other's files, and there is joint annual conference at which the newest developments in the fields of technology, markets, and competition are discussed.

Bureaucratic constraints

In the structured interviews, many CTOs indicated that administrative regulations ('red tape') put clear limitations on their R&D work, even in cases where this was not strictly necessary. The respondents were very critical concerning the number of limitations imposed on the R&D center by administrative regulations (SLCIs 3.6 (2.2) and LLCIs 3.9 (1.7) on the seven-point Likert scale). A critical review of the different regulations could therefore be called for. Some of the companies work with an employee survey system in which the staff is asked to provide their comments and recommendations about technical and management matters. The results are then reported back to the staff and they are asked to prioritize the recommendations. The procedure is conducted externally in a formalized way to ensure total confidentiality.

Internal networks

All participating companies reported in the structured interviews that much had been done in the last few decades to break down the barriers between R&D and the rest of the company by building networks. These networks can either be formal or informal (subject-based, knowledge-based, competency-based or business process-based). They transfer information outside the firm's hierarchical organization and in this way avoid bureaucratic barriers within or between organizations. They are considered to play an important role in the development of key technologies by implementing common methodologies and tools across the organization and coordinating R&D planning across business units, and they may also help to move people and work across organizational boundaries. In nearly all of the participating companies the building of networks was actively encouraged, with the average number of networks being between 20 and 25. Only one company indicated that after a phase of active stimulation of the building of networks, it had arrived streamlining their efforts. In the structured interviews the following reasons for the building of networks were mentioned:

- to increase efficiency by resource optimization;
- to enhance organizational learning by transferring knowledge and technical know-how throughout the company;
- to gain access to competitive intelligence; and
- to gain information about and influence (inter-)national requirement standardization.

In almost all the participating companies the results of networks are monitored carefully and assistance is given in the building of new networks. More than half of the participating companies had an active policy towards the publication of network results, either in hard copy or on the company's intranet. In one of the companies the main emphasis was placed on technical networks. Six such networks are distinguished here: sensors, processing systems, software, display, integrated product development, and systems. Each network included four to five sub-networks, and was coordinated by a Network Executive Committee. Since all the network members had full-time job assignments, each network has a facilitator provided by Corporate Office. Corporate money was provided to cover out-of-pocket expenses such as outside speakers and an annual event was held at which each network presented its latest endeavors.

5.6.4 External communication

Table 5.11 shows the self-assessment of the degree of customer communication in SLCI and LLCI centers.

As proposed, the level of customer orientation was higher in SLCIs than in LLCIs, being closer to the customer. In the SLCI centers project ideas were significantly more consistently evaluated in terms of their value to customers. Also in the structured interviews and in the answers to Questionnaire I it became apparent that SLCI centers put more emphasis on the

Table 5.11. Self-assessment of the level of customer communication, seven-point Likert scales, mean and (s.d.).

	SLCIs	LLCIs
New product forums involving R&D staff held regularly	4.5 (1.8)	3.9 (1.3)
Project ideas evaluated in terms of their value to customers	6.4 (0.5)	5.4 (1.1)**
Structured tools are used to translate customer requirements	5.2(0.6)	5.2 (1.1)

**p < 0.05

direct contact of R&D staff with the external customer: on average 5% to 10% of working time is spent on direct contact. One CTO stated that optimizing customer orientation was the basic philosophy of the company, underlying all its activities. This company reported that its R&D staff spent about 15% of their working time on direct customer contact. The following tools were mentioned in the structured interviews for improving customer orientation:

- customer focus groups to identify customer needs and to react to concepts and prototypes;
- beta site tests to evaluate prototypes and test pre-commercial product performance;
- conjoint analysis to quantify customer trade-offs across different features; and
- placing customers in different scenarios to evaluate future products.

All participating companies put an emphasis on the monitoring of the R&D environment to find new ideas and enlarge their R&D network. However, the interviews indicated that there was a natural tension between the openness which goes with conference attendance and publication and the secrecy needed to protect R&D knowledge. As one CTO put it: *'No industrial secret can stand the fifth glass of wine'*. Seen in this light we were not surprised to find a nearly inverse relationship between the scientific visibility of an R&D center in terms of conference attendance and publication on the one hand, and patent submission on the other. The LLCI company with the lowest number of patents submitted (about 0.2 patents per year) reported the highest conference attendance and numbers of publications. However there was one exception, the SLCI company reporting the highest level of patenting (1.9 patents per 10 R&D employees) also reported the second highest level of conference attendance and publications. Apparently it is possible to combine a high level of scientific visibility with enough secrecy to guarantee patent submission.

In accordance with Proposition P(C)3 about more intensive external scientific communication in LLCI centers, the finding was that in 80% of the R&D centers in LLCIs, scientific publications and presentations at international conferences were actively stimulated. Each staff member in LLCIs attended an average of one international scientific, technical or managerial conference per year, half of them in universities, the other half in industry. In about a quarter of the cases they presented a paper at these conferences. On average one in seven scientists and engineers writes an article in an international scientific or technical journal per year, increasing the scientific visibility of their lab and further motivating the R&D staff involved. In some companies publication targets are included in the remuneration scheme. In one company an article in a refereed journal was awarded US$ 150. The interview indicated that an extensive clearance procedure for external publications ensures that competition-sensitive information is not disclosed and only those journals that do not require the full disclosure of technical details are selected.

Another measure for visibility is the number of visits by scientists and engineers from other companies and scientific institutions to the centers. On average 200 to 250 foreign visitors visited the R&D centers each year. Two well-known centers reported a much higher number of over 500 visitors. Almost all the participating companies emphasized their support for these visits in spite of the burden they place on man-power. Visits result in a positive image for their labs, provide the possibility to attract new young staff from universities and polytechnics, and assist the R&D networking function.

Many of the participating companies spend a lot of time on the eco-efficiency of their products and services and are involved in external communication with the authorities and the general public about environmental issues. One of the companies had even started a biennial award for the best idea in the field of eco-efficiency and sustained development.

5.6.5 Incentive systems

Most of the participating companies were very attractive to young R&D professionals, being 'technology leaders', and having offices in many countries. The R&D centers showed the following 'technology leadership' characteristics: a strong focus on technical excellence; self-organization and bottom-up product and process development; and dominance in an engineering outlook.

There was also a high morale and esprit de corps among the R&D staff in the participating companies. R&D is certainly not a 9-to-5 job. Absenteeism due to illness is low at about 2% on average. One US R&D center reported an even lower absenteeism rate due to illness, namely 0.5%. The CTO stated that he felt that things had gone too far and that too often R&D employees who were obviously ill still came to work.

The CTOs in the European companies indicated that because of their high-tech image, monetary payments to employees could be somewhat lower than in competitor companies. The US-based companies, by contrast, reported having a salary level that was (somewhat) higher than those of competitors. Half of the companies provided a company car and/or had some sort of patent profit-sharing system, and three-quarters provided extra salary for excellent achievements. One of the US-based companies reported having a results-based salary scheme that reinforced both technical excellence and business alignment. The bonus system counted for 5% of the basic salary for average scientists and technicians and up to 40% at the R&D management level. Once a year every R&D staff member got the opportunity to present his/her work to the R&D directors and the lab managers to evaluate its technical excellence. And specific R&D objectives, including dates and performance goals, were formulated on an annual basis and evaluated with the business units. Objectives may be subject to reformulation if business unit needs change, or as a result of unforeseen technical problems or discoveries. Scores are posted where they can be seen by all employees and visitors, so R&D staff members working on lower-scoring projects become motivated to work hard to improve them. Since an outline of recent publications and patents is on display, the work of the lab is made known to other stakeholders, such as universities and contractors. Furthermore, the publicly-posted scores enable administrative personnel, who provide support across many projects, to gain increased knowledge and a proprietary feeling towards the projects they support.

Recognition

A clear difference was found between labs in SLCIs and those in LLCIs concerning their attitudes towards recognition. Although all the respondents indicated that public recognition was an important tool for R&D staff motivation, the companies in LLCIs clearly put more emphasis on it. For instance, all companies in the LLCIs had fellowships and a technical society for excellent researchers. In only 30% of the SLCI centers were such immaterial incentives present. Less frequently but still in 60% of the LLCIs and (again) in 30% of the SLCI centers, special funding to pursue researchers' own 'crazy' (as one respondent put it) R&D projects was provided.

All the participating companies had awards for special contributions, such as plaques in prominent places, photographs and articles in the company's journal, 'Inventor-of-the-Year' schemes and recognition dinners with higher management. One company had 'a staircase of awards' as the respondent called it. The interesting thing is that this company provided awards across the board, not only for the top innovative activities but also for technical and administrative support staff. As one respondent put it:

The special importance of the technical and administrative support staff for the well-functioning of an R&D lab is often overlooked.

There are also different kinds of team recognition activities, for instance, interim and completion dinners for project teams at private, off-site celebrations that bring closure to their efforts and release some of the pressures which have built up during the push for completion. In addition, long-lasting reminders of their contribution, such as pins, trophies and certificates of appreciation, are given.

Recruitment, learning and education

Most of the R&D centers recruited their R&D staff in the center's host country. Only in one of the labs under study was it policy to have a staff composition that reflected the multinational side of the company in order to improve contacts with plants in different countries. Special tax facilities, combined with a special program, including language training, housing and formalities, were in place here.

All participating companies pay great attention to leverage the intellectual capital of their staff by encouraging learning and education. The typical training budget counts for about 10% to 15% of the material budget, which is much higher than in most companies. Each staff member spends on average one to two weeks a year on training, and on average 3% of the R&D staff are away at any one time on training programs. These may include short technical or statistical training programs or courses in more general business skills such as marketing, negotiations and giving oral presentations. The length of a program varies from one day to two weeks to special MBA programs of several years to prepare promising scientists and engineers for management tasks, and in some cases even full-length PhD programs.

5.7 R&D process

5.7.1 R&D resources

The participating companies all had one or more specialized or integrated central research centers where 0.4% to 2.8% of the company's staff worked, with a median of 0.5%. In the structured interviews, several reasons for organizing R&D in a central center were given: to create technical synergy across businesses; to sustain a firm's key technological competencies; and to reduce the duplication of equipment and skills.

Indeed for companies that follow a strategy of technological leadership and are able to capture technical synergies across product lines a centralized R&D center offers a good method of control. It also facilitates cost-effective access to leading-edge equipment and facilities and, as Chester (1994) asserted, a favorable return on R&D investment. Companies that cannot capture sufficient synergy, either because of differences in product lines or because of corporate intent to seek financial and market synergy instead of technical synergy, find conducting R&D within business units a better approach.

If we look at the organizational structure of the labs under study, we see that on average 4% to 5% of the R&D staff were part of the R&D Directorate (in one company even 9%), 28% were involved in research and technology development (one LLCI company reported a much higher percentage of 68% of its R&D staff); 54% were working on product and process development; and the remaining 14% were technical and administrative support staff. A number of interviewees indicated that technical consulting services were just as important for a central R&D center as inventing new or improved products and processes.

5.7.2 R&D throughput

Figure 5.5 shows that in all industries most of the time is spent on product and process engineering. In the participating companies on average half of their time was spent on this phase. It is interesting to note that the participating companies reported spending more than a third of their time on concept development and the specification and planning phases. McGrath (1995) reported that the more time is spent on these early phases in the R&D process, the lower the number of projects that had to be abandoned later.

Governance of R&D projects

Both the R&D development and R&D program managers in SLCIs and LLCIs were positive about the way R&D projects were monitored {5.9 (1.3) in SLCIs versus 5.7 (1.0) in LLCIs on a seven-point Likert scale}. All the companies reported that over 70% of their projects met their original specifications, as well as the needs of the customer. Only one company reported having major problems with staying within the original budget, although (or perhaps because) the reported priority for meeting this goal was not very high.

The participating companies organized their R&D in two ways. Either they installed a limited number of very large projects, typically two to four projects with a large project team of 30 to 60 R&D staff members, or they had a much larger number of teams, 15 to 45 depending on the size of the lab, and a project team size of 3 to 10. In the first category only representatives of the different functions in the project teams met to coordinate activities. The R&D work

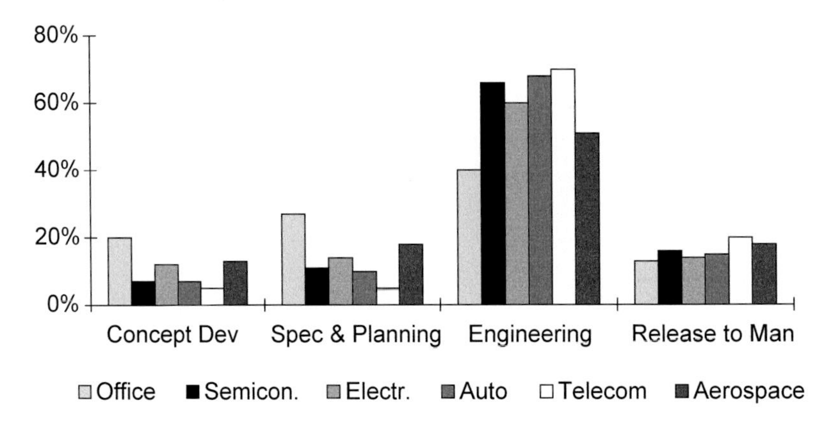

Figure 5.5. Percentage of time spent on the different phases of the R&D process. Sources: the present study and McGrath (1995).

was mainly conducted in the functional departments. In the second category the project teams were much smaller and met regularly. In some companies the team worked in a separate room away from the functional department for the entire duration of the project (the one-room approach).

The disadvantage of the first system is that coordination and communication problems may arise; especially if the contact persons (the linking pins) fail to pass on al the vital information to the R&D staff involve. Gerritsma and Omta (1998) concluded that, above a certain threshold, communication and coordination problems rise almost exponentially with the increase in project size and complexity.

As was indicated in Section 5.4.2 on R&D portfolio planning, most participating companies had trimmed the number of R&D projects they had in the years before the study started. This decision is supported by research that indicates that the best R&D results are attained if engineers work on two major projects at the same time (Liker and Hancock, 1986). When engineers focus on only a single project there is a risk of sub-optimality because they may have to wait for essential information, and when they work on too many projects an increasing fraction of valuable time is spent on non-value adding activities such as coordinating, attending meetings or tracking down information. In the participating companies on average one to three projects were conducted per engineer. Only one company indicated that the project team members were involved in five projects at the same time.

The one-room approach

In the one-room approach a cross-functional team works together in a special facility for the time-span of an R&D project. Following the 'Kaizen' philosophy, in one of the participating companies the walls of the rooms were covered with maps, outlining the most important project measures (the 'glass walls') to enhance cross-functional communication, and as an important source of 'early-warning' information for higher management. From the interviews it became apparent that the greatest advantage of the one-room approach is found in the period of concept development, specification and planning. In that period the need for intense communication between disciplines and functions is the greatest. After this initial period the advantage of improved communication is less and the disadvantage, that valuable R&D staff members are difficult to reach for technical advice and help within the functional setting, generally predominates. There is also the threat that alienation occurs between the 'bright guys' in the 'one-room' and the others working in the functional setting.

Project team organization

Nearly all the companies had full-time project leaders appointed to their major R&D projects and only two companies worked with part-time project leaders. The frequency of project team meetings was weekly or biweekly. One company, with a large project team size, indicated that the project teams only met once every three months. Research by Omta and Van Engelen (1998) showed that if formal communication was rare, the necessary coordination took place via informal channels, which generally proved to be inefficient, especially if functions had to work in parallel. In more than two-thirds of the companies, manufacturing and marketing participated in the project teams on a regular basis. Finance, major customers and suppliers participate in about a third of the companies. The participation of other companies, such as engineering firms, was much less. In only two companies did firms participate in about 20% of the project teams.

Manufacturing and marketing were involved in most R&D projects. The assignment of staff members from finance to the projects was rare, namely in less than 5% of the projects. Purchasing was involved once it was handed over to the divisions. Furthermore, legal experts in patenting or other matters were very important in US projects and some companies were also planning to bring in staff members to better assess safety and environmental issues.

Stage gate review

Most participating companies used some kind of stage gate review system, with typically four or five gates, in which a cross-functional team, including staff members from R&D, marketing and manufacturing systematically reviewed the larger R&D projects. Reasons for termination were too-high development costs and unexpected changes in market and technology development. To make the review process more 'objective' in some companies, external experts were also involved in the gate review system. Such gates were typically at three to six monthly intervals, and at each gate a number of questions were asked. If the answers were positive the project continued. A problem encountered in a number of companies was that more projects should be stopped than actually happened. According to the respondents, vague answers were too often accepted. An additional problem was that essential information for a reliable go/no-go decision was often missing because not all the parallel activities were ready for the stage gate review.

One of the participating companies extended the stage gate review system to include basic and applied research by inserting science-oriented milestones measuring progress vis-à-vis targeted technological advancements and contributions to road map achievements.

A problem that was encountered in some of the participating companies could be called 'responsibility avoidance'. It was stated that this was difficult to change because it was embedded in the R&D culture. In one of the companies a system was in place to address the problem. It involved clear indications as to which person or which level in the organization was responsible for all the key issues and activities.

5.8 Methods to improve strategic alignment

Most of the management measures that can be taken to improve strategic alignment of innovation to business have already been discussed in the previous sections, i.e. R&D projects evaluated in terms of alignment to business; R&D projects prioritized as to customer value; cross-functional participation in R&D planning; R&D project prioritization by BUs; R&D project parameters discussed with BUs; and R&D participation in BU business plan formulation. In addition, there were a number of well-known management tools in use, such as the Quality Function Deployment and DFX, Design for Manufacturing, Service etc. which are not relevant to the discussion here. Virtual development tools, such as CAD/CAM, were increasingly being used to increase the pace of product development. Because they are well known, we will not elaborate on them in this section.. Here we elaborate on the R&D funding structure, collegial network steering and the use of market-technology road maps and staff exchange.

5.8.1 R&D funding structure

Until the 1980s, most central research centers received their financial support directly from corporate office. In the 1990s this pattern changed towards a situation where at least part of the R&D budget was funded directly by a company's business units. By doing this, the locus of control was shifted from the corporate to the BU level. The basic idea behind this shift was to improve the alignment of R&D and the BUs by transforming the latter to external customers. The measure of customer satisfaction was the business units' willingness to fund and to continue to fund specific research projects, leading to business-driven R&D.

Figure 5.6 compares the figures of the participating companies with the funding situation of other technology-based companies. It shows a clear tendency for companies to shift from corporate to BU funding. There were labs that were still totally corporately funded, and there were labs which got all their money from the BUs. On average 50% of total funding came from business units.

In a number of the interviews concern was expressed that the short-term orientation on business needs could gradually result in the deterioration of the technology knowledge base, which is a major source of long-term learning and survival for a company. A CTO of one of the R&D labs with a high percentage of business unit funding expressed this as follows:

This might become a weakness in the long run because the business units tend to be too focused on short-term profit, especially at times when profitability is not too good. This tendency is increased because of the emphasis on shareholder value. They want short-term results on the money they have invested....The time horizon should be longer to reduce the risk of technical surprise and missed opportunities on the one hand and to fulfill the needs of tomorrow's customers, on the other.

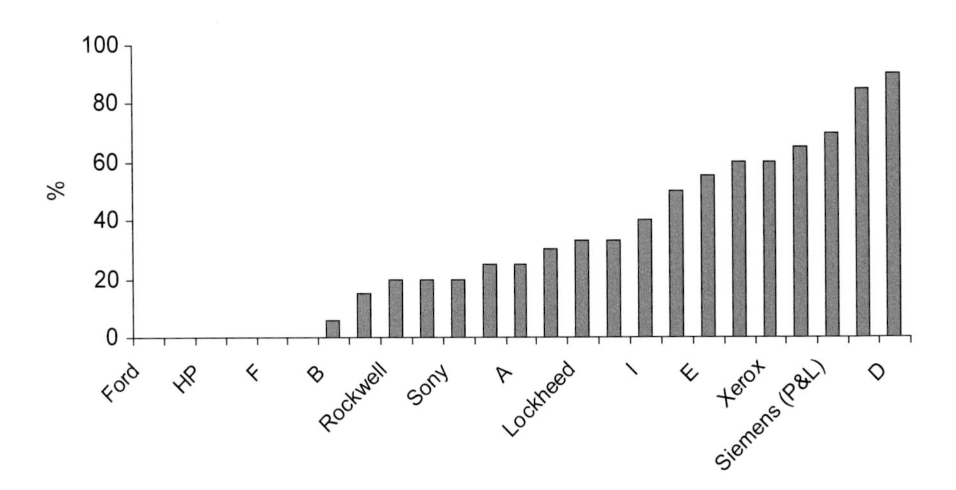

Figure 5.6. Percentage of business unit funding in the participating companies (A to J) compared to other technology-based companies Sources: the present study and McGrath (1995).

To stimulate research with a longer time horizon, a number of companies make use of a hybrid system, in which part (typically a third to a half) of the budget of promising long-term R&D projects stems from corporate R&D funds subject to matching from R&D funds held by the business unit(s). In this system alignment of long-term research to business needs is guaranteed because if business units do not want to pay their share, corporate money is withdrawn. The corporate funding of at least part of the budget guarantees that long-term research receives enough attention.

One of the US-based companies in the present study had even gone one step further by putting the corporate R&D lab at arm's length in a private company, providing it with the chance to work for other companies as well. In effect, the lab would work for at least three, mutually linked, companies. This lab already had a long and positive record of conducting contract research for government agencies. At the time of the site visit, the R&D lab had been autonomous for only a week, so it was far too early to draw conclusions. In the interview, the following expectations about advantages and drawbacks of the new situation were mentioned. On the positive side, these were a stronger customer focus and more freedom to choose the most promising technologies which would lead to faster renewal of capabilities and higher flexibility, and on the negative side, this might be at the cost of a shorter-term orientation.

5.8.2 Collegial network steering

In the system of collegial network steering, managers and researchers from different parts of the organization meet to discuss the best ways to capitalize shared resources. In one of the participating companies, the network consisted of a matrix of 19 cross-disciplinary research domains on the one axis and the research labs in the different businesses of the company on the other. These 19 domains included 250 topics and 700 different pools of expert capabilities. Basically, everybody in the company had access to a capability database of experts in all businesses. The 250 topics were selected to reflect real industrial needs and technological evolution. Those experts, who worked on a specific domain, gathered to discuss themes of common interest. In addition, every month the technical directors of all the research labs met at the headquarters to discuss the steering of the network and to resolve problems. Typically, two-thirds of the meeting was spent on management topics, for instance mutual internships and the coordination of multi-site projects, and only a third of the time was spent discussing and solving technical problems. Every three months this committee reported to the Executive Committee.

The system turned out to work well and coordinated action was taken more easily. In the past, many experts did not feel valued by headquarters, and in a diversified conglomerate, they could not see the results of their work. There was also overlap, especially in applied research, across the businesses. There are numerous problems with collegial steering. Each company has its own culture, so the frequency of meetings must be chosen carefully. These should be compulsory so that everybody attends but not too frequent. There must be enough topics of common interest so that attendance does not feel like a burden. However, the clear structure, deriving from the strict definition of the research domains in this study, turned out to be a critical success factor.

5.8.3 Market-technology road mapping

The purpose of market-technology road mapping is to identify which future skills and technologies are needed to meet longer-term business needs. A typical market-technology road map looks from five to ten years into the future and is based on marketing studies of customer requirements and competitor activities that are integrated into the desired product and process plans.

Most of the participating companies made use of market-technology road maps but not consistently, especially in LLCIs. Market-technology road mapping, for instance, is not integrated in sales expectations. However, a CTO of a company which made extensive use of market-technology road maps indicated that the market-technology road map proved to be one of the most powerful tools for integrating technology and business strategy. It helped his company to solve one of its major R&D problems, namely the time gap between invention and application. It made transparent to businesses how technologies in the past contributed to their current profitability, and likewise, how emerging technologies could contribute to their future profitability.

In addition, the road map provided a time-table, and showed everyone in R&D what, and how large, the expected improvements were for product generations to come. In the past,

R&D staff was allowed to work on small improvements in customer value. Now that future customer needs were mapped, a greater part of the R&D effort could be directed towards major improvements in customer value. Furthermore, the road map forces R&D to consider upfront whether a new technology should be developed in-house, or if it has to be found elsewhere, either outsourced, co-developed or in-licensed.

5.8.4 Staff exchange

Staff exchange was encouraged in most of the participating companies. Internal vacancies were posted and outside experience (e.g. fellowships) was considered in senior level promotions. In the interviews it was indicated that the R&D management should fulfill a coaching role by enhancing the international exposure of their staff. With the increased cycle time reduction sharing engineers among R&D labs had become easier because they were away for a shorter time. This offered new challenges because the staff in the recipient lab often regarded the newcomer as a transient who would stay for a limited time period only. A careful introduction is important. Company-wide joint human capital ICT platforms showing the capabilities of all experts may facilitate the reception of staff in the recipient organizations.

However, many of the participating companies showed the 'technology leader' characteristic in that most engineers started in R&D and either stayed there or moved up to the divisions. Despite the attention given to staff exchange there was a general feeling that the number of annual staff transfers from R&D to other functions in the organization and vice versa was too low. In the interviews it was indicated that many R&D staff members did not want to move for personal reasons, while some were very attached to their research projects. Remarkably, this feeling was not reflected in the reported annual staff flow, either to other departments or to other organizations, which typically varied from 0.01% to 10% (!) of the R&D staff.

5.9 Concluding remarks

In this chapter the results, obtained from the cross-industry study are presented and confronted with the propositions, formulated in Section 4.3.4. Table 5.12 provides an overview of the propositions and the empirical results. We conclude that clear support is found for a number of the propositions (see Section 4.3.4). The PGLC can indeed be regarded as a relevant indicator for classifying industries in short and long life cycle industries, since the empirical results show definite differences in the proposed directions between SLCIs and LLCIs. With regard to our first research question, we therefore conclude that the market and technology forces of the business environment do indeed affect strategic alignment, and that their effect can be predicted when the length of the PGLC in a given industry is used as an indicator of the combined effect of these forces.

In accordance with P(C)1, strategic alignment is clearly perceived better in SLCIs than in LLCIs, both in terms of internal as well as external fit. In accordance with P(C)2, the R&D

Table 5.12. Confrontation of proposed differences between short life cycle industries (SLCIs) and long life cycle industries (LLCIs) with the empirical results.

Research variables	Operationalizations and indicators	Predicted		Results	
		SLCIs	LLCIs	SLCIs	LLCIs
Strategic alignment					
External fit	Market and technology trends imp. Inputs	+[1]		++[2]	
Internal fit	BU point of view on meeting business goals	+		+[3]	
R&D strategy					
R&D performance	Sales revenues from new products	+		++	
	Number of patents per US$ 10 m		+	**++**[4]	
Degree of exploration	Perc. basic research in total R&D; perc. of PhDs		+		++
	Senior man. involved in early development		+		+
	Importance of value engineering	+		+	
R&D competencies					
In-house	R&D competencies monitoring	+		+	
Open innovation	Collaboration with knowledge institutions		+		+
R&D capabilities					
Flexibility: timeliness	Percentage of cycle time reduction	+			**++**
Flexibility: responsiveness	Ease of incorporation of BU requests in R&D	+		[5]	
Internal comm.	Communication with BUs, marketing & sales	+		+	
External comm.	Customer communication	+		+	
	Expert and conference communication		+		+

[1]+ indicates a higher proposed level;
[2]++ strong confirmation: all differences are significant;
[3]+ confirmation: all (or most) of the differences are in the right direction, at least one is significant;
[4]++ (bold) strong rejection: the difference is opposite to the proposed direction;
[5]No significant difference was found.

resources in LLCIs, in terms of emphasis on basic research and the percentage of PhDs among the R&D staff, reflect a more exploration-oriented R&D strategy with a longer-term technology outlook. However, in clear contrast to the proposition, the number of patents per R&D investment is six times higher in SLCIs than in LLCIs. This unexpectedly high number of patents per R&D investment may reflect the essential position that industrial property rights has gradually attained in SLCIs. Also in accordance with P(C)2, a more exploitation-oriented R&D strategy in SLCIs is found to bring new products on to the market regularly, with a focus on value engineering. However, the qualitative results are mixed. The results regarding the R&D competencies are in accordance with P(C)3, where SLCIs concentrate on optimizing in-house R&D competencies to introduce new products on to the market on time. The LLCI companies concentrate more on open innovation with research firms and Research Technology Laboratories (RTIs). However, the findings regarding the R&D capability 'flexibility' contrast with the expectation in P(C)4. LLCIs reported a much higher level of cycle time reduction than SLCIs in the preceding three years, and no significant differences were found in the ease of incorporation of BU requests into the corporate R&D portfolio. Perhaps the unexpected finding regarding cycle time reduction can be explained as catching up by the R&D centers in LLCIs with the R&D centers in SLCIs that had already achieved major improvements. The results regarding the R&D capability of 'communication' clearly support Proposition P(C)4. Both internal communication with customer-oriented staff as well as the external communication with the customer is more intensive in SLCIs, whereas in LLCIs more external communication with experts and at conferences could be observed. We will elaborate later on these findings and their implications for the conceptual framework, and a further interpretation of the unexpected findings will be provided in Chapter 7.

The clearest differences found between SLCIs and LLCIs turned out to be the difference in the strategic alignment situation and in the R&D capability of 'communication'. These research variables were central in the longitudinal study where it was investigated whether and to what extent structured feedback within the R&D/BU/HQ triangle could improve strategic alignment. A number of management tools to improve strategic alignment, namely (partial) funding of corporate R&D by business units, market-technology road maps and virtual development tools were being used by the participating companies. The long-term effect of these technical and management methods on strategic alignment will be investigated in Chapter 6.

6. Longitudinal study results[3]

This chapter reports on the results of the longitudinal study that was set up to answer the second research question.

RQ2. How can strategic alignment of innovation to business be achieved and maintained over time?

In Section 6.1, a baseline description is given of the company and the respondents. In Section 6.2 we take a closer look at the reliability and validity of the reflective research variables using factor analysis, and by comparing early and late respondents. Section 6.3 compares the self-assessment by corporate R&D staff with the assessments of the BUs and headquarters. In Section 6.4 a stepwise linear regression analysis is presented, showing the factors related to external fit. In Section 6.5 the longitudinal analyses of the different research variables are shown by presenting the linear and second order polynomial trend approximations over the six-year period of the investigation. Finally, Section 6.6 provides the reader with some concluding remarks.

6.1 Baseline description

The company employs about 30,000 employees worldwide at approximately 80 production sites in 25 countries. The annual sales volume in 2002 amounted to about US$ 5 billion, with an operating profit margin of about 8% (see Table 6.1). In 1997, 1998, 2000 and 2002 questionnaires were sent to corporate headquarters, the scientific staff of the corporate R&D laboratory and the higher management in the business units.

Table 6.1. Company profile in 2002.

Number of employees worldwide	30,000
Production sites	80
Countries	25
Strategic business units	6
Annual sales volume (US$)	5 bn
Operating profit margin	8%

[3] Three papers were published reporting on the longitudinal study: Fortuin and Omta 2007b, Fortuin and Omta, 2006 and Fortuin *et al.*, 2005.

6.2 Study sample, reliability and validity

The total study sample consisted of 696 respondents, 213 from the corporate R&D centre (69 in 1997, 67 in 1998, 44 in 2000 and 33 in 2002) and 483 (147 in 1997, 189 in 1998, 102 in 200 and 45 in 2002) from headquarters and the strategic business units. This included 83 directors, 253 product and process managers, 91 marketeers and sales managers and 56 with other functions (e.g. engineers). The average response rate was 67% for the corporate R&D staff and 44% for the HQ/BU staff, although the response rate was clearly going down during the investigation period, probably because of questionnaire fatigues. If we look at the number of times that the respondents from the corporate R&D center on the one hand, and those from the BU and headquarters on the other participated, we see that 68% of the R&D staff participated only once, 16% twice, 11% three times, while 5% participated all four times. In the BUs and headquarters the numbers are as follows: 70% participated once, 21% twice, 7% three times, and 2% participated in all four surveys. An important explanation for this finding is the high mobility rate of the company's staff. If we focus at the more time respondents, we see that many of them has changed position, from one BU to another, even from R&D to the BUs and vice versa, and also the positions within R&D and the BUs change. One of the BU respondents that participated in all four surveys, started as an engineer in 1997, had become BU manager in 2000, and ended as the director of another BU in 2002.

As stated in Section 4.4.3, to correct for the effect that respondents who participated more often might tend to give a more positive judgment based on the feedback of the earlier survey, the analyses were checked using the independent sample of the total database (n=474), meaning that every respondent was entered only once. Comparing these results with the results obtained using the whole database did not change the conclusions. To establish the representativeness of the study sample, the answers of early, average and late respondents were compared, because Oppenheim (1966) suggested that late respondents resemble non-respondents rather than early respondents.

Since no differences were found in the answering patterns of early, average and late responders, the sample is considered to be representative for the study population as a whole. Also, no significant differences in opinions regarding R&D competencies, capabilities and alignment of the corporate R&D center were found between the different BU functions. Only, interestingly, the marketeers and sales managers were significantly more positive about the competencies of the corporate R&D center ($p < 0.05$) than the other respondents. Because of their limited numbers (3 to 5 per survey), respondents from corporate headquarters are ranked among those of the directors function of the business units.

Table 6.2 presents the factor structure, and only the factors with an 'eigenvalue' above 1.0 and the items with a factor loading above 0.4 are presented. The factors are listed in the order of presentation in the variable lists, and the factors are named in accordance with the names of the corresponding research variables. Comparison of Table 6.2 with Table 4.2 shows that the

Table 6.2. Factor structure and Cronbach α of the reflective research variables. Principle component analysis with Varimax rotation (n=696).

Factors	Factor loadings
F1 Strategic alignment, external fit (eigenvalue 4.30, explained variance 23.9%, Cronbach α 0.85)	
Technology Board-funded R&D projects concentrate on important technologies	0.66
Technology Board-funded R&D projects align with market needs	0.87
BU-funded R&D projects concentrate on important technologies	0.75
BU-funded R&D projects align with market needs	0.86
F2 Responsiveness (1.17, 6.5%, 0.72)	
Ease of incorporation of BU requests into R&D portfolio	0.83
R&D start-up time lag	0.78
F3 Timeliness (1.81, 10.1%, 0.68)	
Cycle time of R&D projects	0.75
R&D projects delivered on promised date	0.79
Time-lag to answer technical questions	0.59
F4 Communication (2.02, 11.2%, 0.65)	
R&D-BU contact during project execution	0.73
R&D-BU contact after project completion	0.76
R&D – BU Contact after new products/processes are introduced	0.64
Clear reporting of R&D project results to BUs	0.44
Serious analysis of BU complaints	0.47
Communication importance (Cronbach α 0.64)	
F5 Internally (1.13, 6.3%)	
Importance of regular R&D-BU meetings to assess future needs	0.74
Importance of regular R&D-BU staff exchange	0.76
F6 Externally (1.28, 7.1%)	
Importance of regular R&D–end-user meetings to assess future needs	0.88
Importance of regular R&D–end-user meetings to assess quality	0.90

factor structure confirms the predefined structure of the research variables. Table 6.2 further shows that in all cases Cronbach α is sufficiently high (> 0.64) to warrant confidence in the internal consistency of the scales in exploratory research (Hair *et al.*, 1998).

6.3 Comparison of R&D and BU/headquarters assessments

Table 6.3 shows that the self-assessment of the R&D staff concerning R&D competencies, R&D capabilities of timeliness and responsiveness, and strategic alignment is significantly

Table 6.3. Comparison of R&D self-assessment with the assessments by BUs and headquarters. t-tests, mean and standard deviation (between parentheses, n=213 for R&D and 483 for BUs).

Research variables	**R&D**	**BUs/HQs**
Strategic alignment: External fit	4.90 (1.02)	4.15 (1.17)***
R&D competencies	3.80 (0.55)	3.52 (0.54)***
R&D capabilities: Responsiveness	4.13 (1.43)	3.84 (1.49)*
R&D capabilities: Timeliness	4.62 (1.23)	3.93 (1.23)***
R&D capabilities: Communication	4.49 (0.84)	4.42 (0.91)
Communication importance: Internally	6.23 (0.85)	5.98 (0.97)**
Communication importance: Externally	5.54 (1.34)	5.08 (1.69)**

* $p < 0.05$;
**$p < 0.01$;
***$p = 0.000$.

more positive than the assessment by BU respondents and those from corporate headquarters. Only in the actual communication situation were no significant differences found between the two types of respondents. But significant differences occur if we look at the methods for improving communication.

This is to be expected: outsiders are generally more critical about the way certain activities are conducted then the people who actually perform them. It should be realized that these data are averages over the whole six-year period of investigation, Table 6.6 and the longitudinal graphs presented later in this chapter show that these perception gaps diminished over time for the most important research variables.

Table 6.4 shows an interesting contrast in the Pearson correlation matrix of the corporate R&D center and that of the BU and headquarters respondents. We interpret these correlations as follows.

The finding that in the perception of the R&D staff strategic alignment and R&D competencies are negatively correlated with communication and the importance of internal communication seems odd at first sight. We assume that from the R&D staffs' point of view, there is consistency in it, the higher their self assessment of the achieved level of strategic alignment and their own technical skills and know-how, the less importance is attached to further improving the communication with the business units.

The finding that all communication variables are significantly positively correlated with external fit in the BU and headquarters, supports the expectation that the capability of communication

Table 6.4. Pearson correlation matrix of the self-assessment by corporate R&D staff (n=213) versus the assessment of the respondents from BUs and headquarters (n=483).

Research variables	R&D comp.	Resp.	Timeliness	Comm.	Comm. imp. int.	Comm. imp. ext.
Corp. R&D center						
R&D competencies	X					
Responsiveness	0.19**	X				
Timeliness	0.14*	0.19**	X			
Communication	0.19**	0.08	0.13	X		
Communication imp.: internally	-0.09	0.01	0.01	0.01	X	
Communication imp.: externally	0.09	-0.16*	0.01	0.18*	0.22*	X
Strategic alignment: external fit	0.34***	0.18**	0.36***	0.16*	-0.16*	0.01
BUs/headquarters						
R&D competencies	X					
Responsiveness	0.14**	X				
Timeliness	0.26***	0.45***	X			
Communication	0.36***	0.28***	0.33***	X		
Communication imp.: internally	0.14**	0.01	0.05	0.04	X	
Communication imp.: externally	0.17***	-0.04	0.03	0.11*	0.26***	X
Strategic alignment: external fit	0.39***	0.35***	0.37***	0.30***	0.20***	0.11*

*$p < 0.05$;
**$p < 0.001$;
***$p = 0.000$.

will be an important factor affecting strategic alignment. The finding that the importance of internal communication is also significantly positively correlated with R&D competencies and the importance of external communication, and that the importance of external communication is significantly positively correlated with responsiveness, is interpreted as a signal, that the more BU respondents have confidence in the technical skills and know-how of the corporate R&D center, and the more they are content about the alertness with which R&D reacts to BU proposals, the more they support the importance of open communication channels with the internal as well as the external customers.

6.4 Factors related to strategic alignment (external fit)

A stepwise linear regression analysis was conducted with strategic alignment as the dependent variable and the R&D competencies, and the R&D capabilities as the independent variables. The following linear regression function was found.

Alignment = 0.42 + 0.98 x R&D funding structure + 0.53 x R&D competencies + 0.17 x Timeliness + 0.41 x R&D or BU/HQ Respondent + 0.15 x Communication + 0.28 x BU/HQ director + 0.05 x Responsiveness

Alignment; R&D competencies; Timeliness; Communication; Responsiveness: scales 1 to 7; Management methods and control variables in regression analysis: R&D funding structure: Corporate funding = 0; BU/technology board funding = 1; Respondent from R&D or BU/HQ: R&D = 1; BU/HQ = 0; BU/HQ director: BU/HQ director = 1; other positions = 0

Besides the R&D competencies and R&D capabilities, the change in governance structure, which was introduced by corporate headquarters between the second and third measurements, came out as the most important management method to improve alignment. The control variable (1) whether the respondent came from the corporate R&D center or from the business units or headquarters; and (5) which position (director, manager or engineer) came out significantly, as well. We will elaborate on these findings in the discussion of Table 6.5.

Table 6.5. Stepwise linear regression of strategic alignment (external fit, n=696).

Research variables	**Beta**[1]	**t value**
R&D competencies	0.26	7.83***
R&D capabilities		
Responsiveness	0.07	2.01*
Timeliness	0.18	5.15***
Communication	0.11	3.33**
Management methods		
R&D funding structure	0.40	12.63***
Control variables		
Respondent from R&D or BU/HQ	0.17	5.15***
BU director	0.08	2.51*
Constant	-	3.19**
Adjusted R^2	0.46	74.27***[2]

*p < 0.05;
**p < 0.01;
***p = 0.000.
[1]standardized beta;
[2]= F value,
Year, function, country, BU, R&D size, integrated virtual development tools, and internal and external communication importance did not come out significantly in the regression analysis.

Table 6.5 shows that 46% of the total variance of strategic alignment (external fit) can be explained by the competency and capability variables, the change in R&D funding structure and a number of control variables. It shows that the R&D competencies (carrying out the right research and doing this in the right way) in the six-year period of investigation became much more aligned to the BU needs. From the R&D capabilities, timeliness of R&D projects comes out as the most important, not surprising if we think of the great decrease in project cycle time that was achieved by the LLCI companies in general (see Figure 5.4). The change in governance structure by shifting the locus of control of the R&D portfolio from R&D exclusively to a joint responsibility of R&D and the BUs and headquarters emerges as the most significant factor determining strategic alignment. The R&D capabilities of communication and responsiveness come out significantly as well. Interestingly, neither the methods to improve internal communication, such as staff exchange, nor the methods to improve external communication with end-users, appear in the stepwise linear regression analysis. The explanation for this might be that while the R&D competencies and capabilities demanded factual information, in these cases an opinion was requested regarding the possibilities of enhancing communication.

The most important control variable is whether the respondent comes from the corporate R&D center or from the BUs. As we will show in the longitudinal trend analysis, it is especially the respondents from the business units who were most positive about the level of strategic alignment over the years, so their judgments contributed more than those of the R&D staff to the changes in perceived strategic alignment. The second control variable that comes out significantly is that of the directors in the business units and headquarters. They changed their opinions on the strategic alignment of R&D to business from a more negative assessment at the start of the investigation to a more positive one at the end compared to the other respondents from the business units (data not shown).

6.5 The longitudinal analyses

For the clarity of presentation of the longitudinal development of the research variables in Table 6.6 and in the Figures 6.1 to 6.7, from each construct that variable was chosen that was central according to the factor analysis (see Table 6.2) and best reflected the issue at stake. Table 6.6 and the Figures 6.1 to 6.7 show that the BU respondents, in particular, became more positive about the corporate R&D center. This assessment increased significantly for strategic alignment (external fit), responsiveness, timeliness (although here the self perception of the R&D staff increased even more) and internal communication improvement. But also the opinion about R&D competencies and communication (after 1998) went up. In the case of external fit and responsiveness, the average assessment of the BU and headquarter respondents was even higher than that of the R&D staff at the end of the six-year period.

From the start, the BU/HQ staff was more positive about the level of R&D-BU communication. After 1998, the R&D staff perception went down, while the BU/HQ perception stayed

Table 6.6 Comparison of the R&D and the BUs and headquarters assessments from 1997 to 2002.

	1997	1998	2000	2002	F value
Strategic alignment: external fit					
R&D center	4.03 (1.53)	4.17 (1.42)	5.59 (1.02)	5.55 (1.03)	21.34***
BUs and HQs	3.33 (1.31)	3.49 (1.34)	5.14 (1.09)	5.64 (1.00)	76.29***
Perception gap	0.70**	0.68**	0.45**	0.09 (-)	
R&D competencies					
R&D center	3.70 (0.44)	3.89 (0.57)	3.79 (0.59)	3.84 (0.62)	1.23
BUs and HQs	3.45 (0.61)	3.51 (0.53)	3.57 (0.47)	3.63 (0.42)	1.37
Perception gap	0.25**	0.38***	0.22*	0.21	
Responsiveness					
R&D center	4.25 (1.53)	4.75 (1.47)	4.58 (1.55)	4.58 (1.80)	1.17
BUs and HQs	3.87 (1.66)	3.96 (1.60)	4.49 (1.54)	4.80 (1,85)	5.95**
Perception gap	0.38	0.79***	0.09	0.22 (-)	
Timeliness					
R&D center	3.50 (1.78)	4.33 (1.64)	5.34 (1.48)	5.21 (1.52)	14.48***
BUs and HQs	2.89 (1.25)	3.40 (1.40)	3.78 (1.65)	3.89 (1.57)	9.80***
Perception gap	0.61**	0.93***	1.56***	1.32***	
Communication					
R&D center	3.78 (1.34)	3.77 (1.43)	3.30 (1.23)	3.00 (1.17)	3.64*
BUs and HQs	4.44 (1.22)	3.94 (1.33)	4.02 (1.32)	4.22 (1.44)	3.99**
Perception gap	0.66** (-)	0.17 (-)	0.72** (-)	1.22*** (-)	
Internal communication importance					
R&D center	5.60 (1.50)	5.98 (1.14)	5.82 (1.19)	6.30 (0.88)	2.57
BUs and HQs	5.13 (1.59)	5.95 (1.08)	6.03 (1.12)	6.09 (1.24)	15.50***
Perception gap	0.47*	0.03	0.21 (-)	0.21	
External communication importance					
R&D center	5.97 (1.16)	5.82 (1.41)	5.73 (1.17)	5.97 (1.10)	0.45
BUs and HQs	5.18 (1.75)	5.18 (1.75)	5.29 (1.74)	4.64 (1.89)	1.50
Perception gap	0.79***	0.64**	0.44	1.33***	

*t-tests + One way Anova; *** p = 0.000; ** p < 0.01; * p < 0.05*

constant. The opposite was true for the importance of external communication. Here the BU and headquarters respondents became more negative over time while the R&D staff's assessment stayed at a very high level.

In addition, the perception gap between the staff of the corporate R&D center on the one hand, and the BUs and headquarters on the other (t-tests), diminished significantly over time for strategic alignment (external fit). This also holds, although to a lower extent, for the R&D competencies and the R&D capability of responsiveness. These results show that for most research variables, internal fit has clearly improved. For the R&D capability of timeliness, the perception gap did not diminish because although the BUs became more positive over time, the R&D staff's self-assessment improved even more. We elaborate on these findings in the next section.

6.5.1 Strategic alignment, external fit

Figure 6.1 shows that the BU assessment concerning the strategic alignment of the R&D projects improved considerably over time.

In 1997 there was a considerable gap between the BU assessment and the R&D staff self-assessment, but this gap has totally disappeared by 2002. In 1997 the BU assessment of the strategic alignment of the R&D projects was quite negative (3.33 on a seven-point scale). This was expected because, as stated in Chapter 1, there often appears to be a certain tension between R&D and the business units, the long-term orientation needed for exploration, and the uncertainty of the outcome being at odds with the predictability needed for executing day-to-day activities efficiently (Roberts, 1995; Glass *et al.*, 2003; Park and Gil, 2006). This initial tension was also in two typical BU statements in 1997:

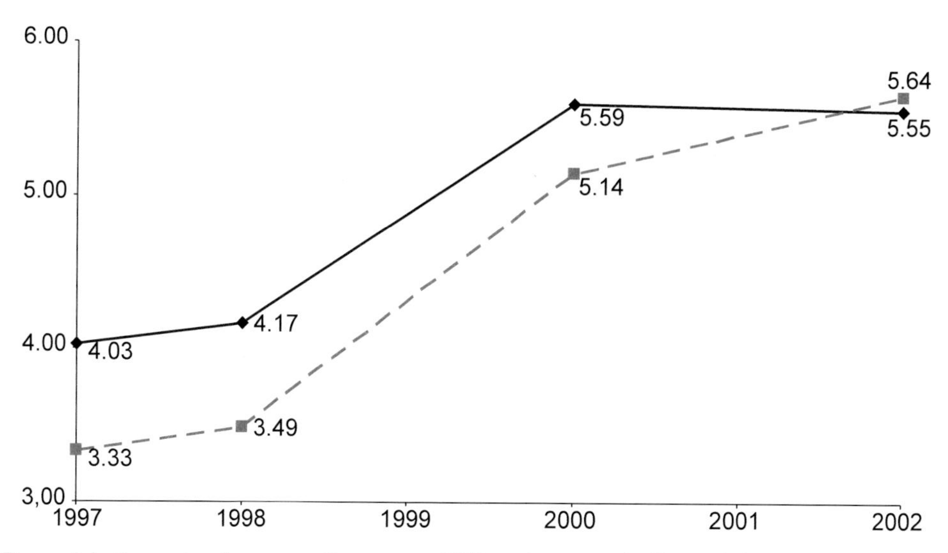

Figure 6.1. Strategic alignment of corporate R&D to business. In Figure 6.1 to 6.7: Continuous line and rhombic points = the perception of the corporate R&D staff; dotted line and square points = perception of BUs/HQ.

The corporate R&D center has its head in the clouds.

Only theoretical studies come from R&D. We need R&D to solve technical problems, never to prepare for the future.

There was also concern about the R&D project portfolio.

The R&D resources are spread over too many projects.

In 2002 the BU assessment was a bit higher than the corporate R&D staff's self-assessment. Typical BU director assessments in 2002 underline this fundamental change in attitude towards the corporate R&D center.

It is important that R&D continues to develop the basic competences in contact mechanics and system dynamics. This knowledge is unique and can provide great value to internal and external customers. Our customers clearly value the work of corporate R&D and see it as the technology core of our company.

Figure 6.1 shows that the BU assessment clearly rose in each successive measurement but the biggest rise can be observed between 1998 and 2000, after the change in governance structure from 100% corporate to a mixed system of 50% business unit and 50% Technology Board-funding took effect. Respondents from headquarters and the BU directors were very critical in 1997, seeing R&D predominantly as a drain on company resources, and they were the ones who had to pay the bill. As one of the BU directors stated in 1997:

Except for one positive case the corporate R&D center has only been a heavy cost burden in the last five years.

In 2002 the situation changed dramatically. Now the assessment of the BU directors and headquarters was even higher than the average assessment.

6.5.2 R&D competencies

As explained in Chapter 4, the variable R&D competencies were composed of two elements: an assessment of the importance of different R&D objectives and an assessment of the R&D laboratory's perceived performance on each of these objectives. Figure 6.2 combines these data in the overall competency level. The longitudinal data show steady progress in the laboratory's overall competency level as perceived by the BUs and headquarters, whereas the R&D staff self-assessment remained constant. This means, that after four successive surveys, the gap between the BU assessment and the R&D staff self-assessment had gradually disappeared.

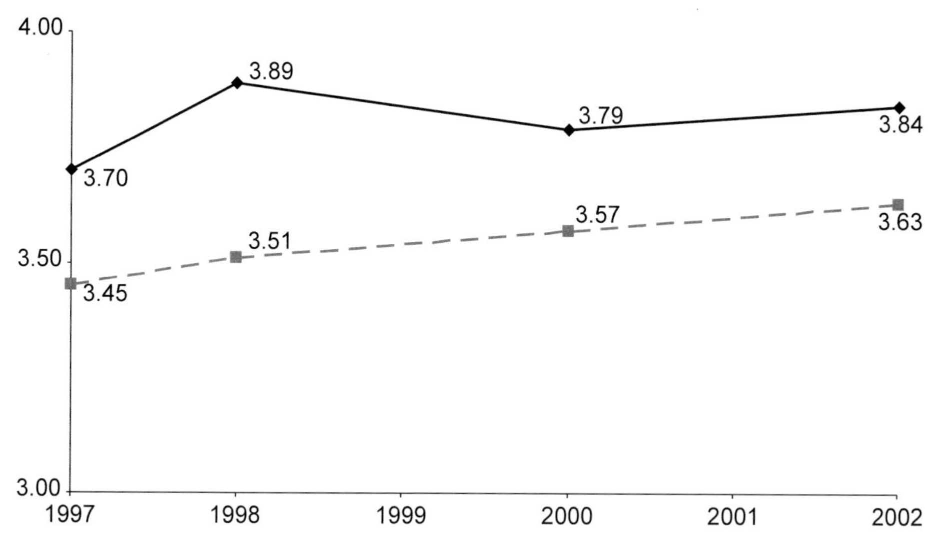

Figure 6.2. Competency level of corporate R&D. Continuous line = R&D; dotted line = BUs/HQ.

6.5.3 R&D capabilities

Flexibility: R&D responsiveness

The results for R&D responsiveness show a similar pattern to those for strategic alignment and R&D competencies, namely a clear tendency for rising Business Unit assessment, and a closing of the gap between the BU and R&D assessments. In the ease of incorporation of BU projects in the corporate R&D portfolio shown in Figure 6.3 the gap between the BU and the R&D assessments has even reversed, indicating that the BU assessment became higher than the R&D staff self-assessment. In 1997 there was clearly room for improvement, as one of the BU managers stated:

The R&D project approval procedure has become abstract and far from the 'basis'.

In 2002, it was much easier for BU customers to have their projects incorporated into the R&D portfolio.

Flexibility: timeliness

The results on timeliness show a positive but much weaker positive trend in BU perception than the former aspects. Figure 6.4 shows the assessment of R&D project cycle time. The reduction of the R&D cycle time is clearly a priority for the company. As a BU Director stated in 1998:

The main objective should be to increase the speed of product development.

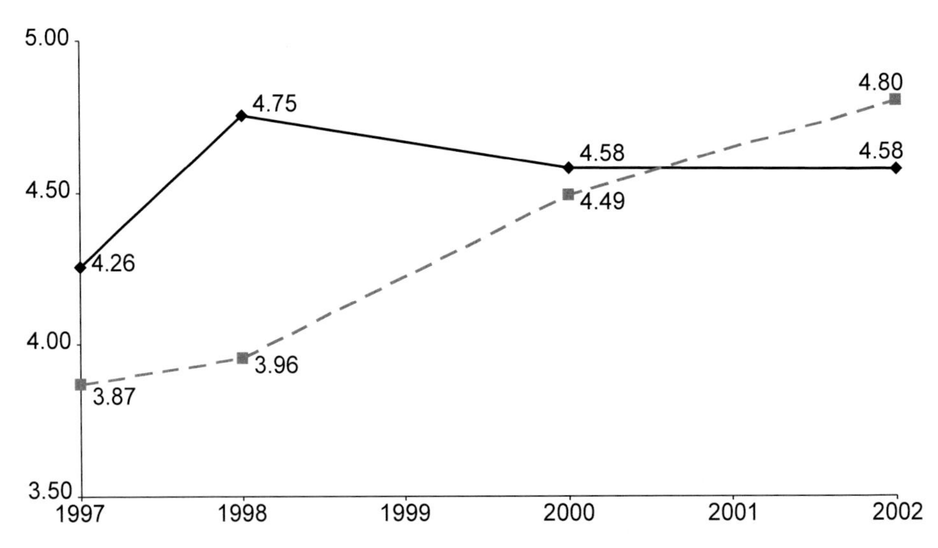

Figure 6.3. R&D responsiveness, ease of incorporation of R&D projects. Continuous line = R&D; dotted line = BUs/HQ.

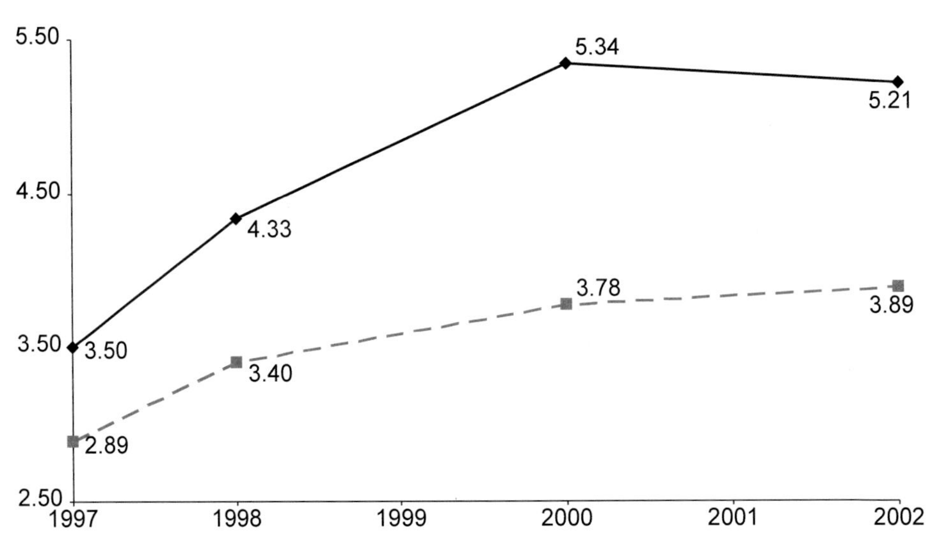

Figure 6.4. Timeliness of project execution, R&D cycle time. Continuous line = R&D; dotted line = BUs/HQ.

The trend in the Business Units' assessment clearly indicates that the feedback has had positive effects but that improvement is only moderate. The gap between the BU and R&D assessment, however, has become wider over time, which was caused by the fact that the self-assessment of R&D staff has risen more strongly than that of the BU customers. A possible explanation for this unexpected finding is that after the first survey, R&D management put a lot of effort into improving R&D timeliness by introducing a Balanced Score Card for R&D (see Table 6.7). R&D staff probably expected the business units to appreciate their efforts but the business units apparently just looked at the results.

Communication

The results of the variable communication show an unexpected tendency. During the whole period of the investigation the BU assessment of the item evaluation of R&D projects results after completion was higher than the self-assessment of the R&D staff and remains more or less stable at a level of 4 to 4.5, which indicates a fairly positive judgment on a seven-point scale. The R&D staff-self assessment staff started lower and declined steadily after 1998. We think this result can be attributed to the fact that the total number of employees in the company, including the R&D staff, was reduced in the period under study. For R&D this meant that they could no longer provide the business units with regular detailed update reports on every project and expected that this would be perceived negatively by their customers in the business units. Apparently the customers did not feel this, thereby indicating that good is good enough.

Table 6.7. Examples of performance indicators in the Balanced R&D Scorecard.

BU Customers
BU satisfaction score
Timeliness (% of projects delivered on time)
Percentage of BU funding in the total R&D budget
Average R&D man-hour costs
R&D Employees
Work satisfaction score
Absenteeism due to illness versus absenteeism due to training
R&D/BU employee transfer rate
R&D process
Budget fulfillment rate, % of the budget coming from external funding (e.g. the EU and external contractors)
Productive hours, the numbers of man hours directly spent on R&D projects versus the indirect hours (e.g. training, study)
Number of inventions proposed and number of patents filed
Number of process improvements in the plants

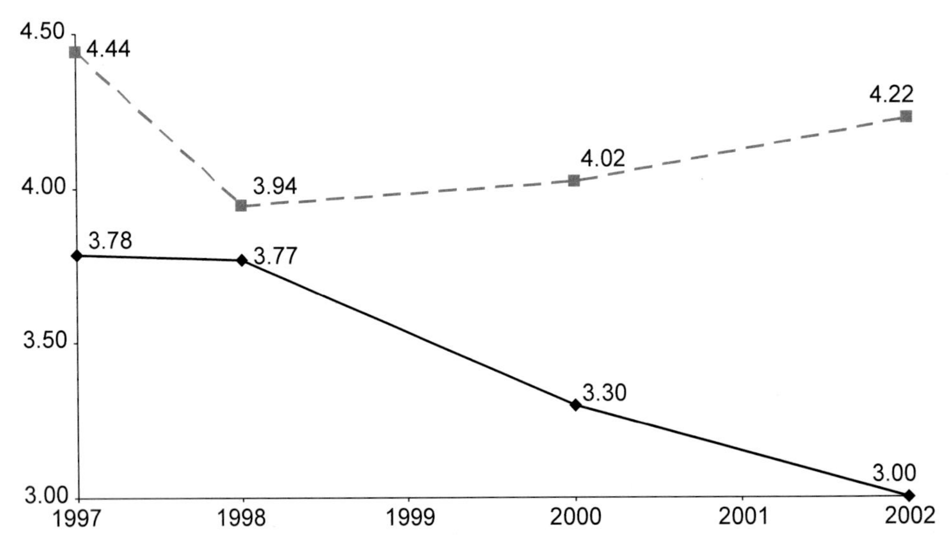

Figure 6.5. Communication, evaluation of R&D project results after completion. Continuous line = R&D; dotted line = BUs/HQ.

6.5.4 Measures to improve alignment

The questions assessing the importance of direct contact between R&D staff and both internal (BU) and external (end-user) customers show two distinct trends: the divisions clearly value regular contact with the R&D center and their opinion on staff exchange as a means to foster communication has improved over the years (see Figure 6.6). In 1997, one of the managers of the BUs pointed at the importance of R&D staff being in the factories.

It would be a good idea if the R&D staff spent some time in the factory, following some phase of the project

A suggestion by a BU manager in 1998 was:

R&D project teams should be hand picked from factories around the world for a period of six months to two years, supported by R&D experts. This way a shared cost base is reached, and people rotate between R&D projects and factory tasks.

In contrast to this, Figure 6.7 shows that in the case of direct communication between the R&D staff and the end-users, the gap between the R&D center and the BUs/HQ widens over time. The BUs/HQ are clearly not in favor of the idea of R&D staff having regular contact with end-users independent of the Business Units. Their main fear was that R&D staff would offer solutions to end-users before a commercial price could be negotiated. As one of the BU directors stated in 1998:

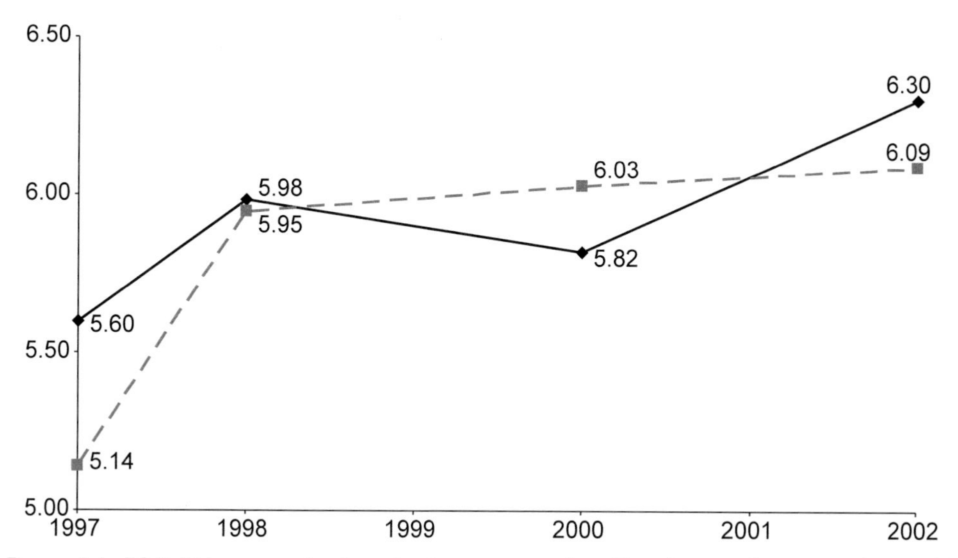

Figure 6.6. R&D-BU communication, the importance of staff exchange. Continuous line = R&D; dotted line = BUs/HQ.

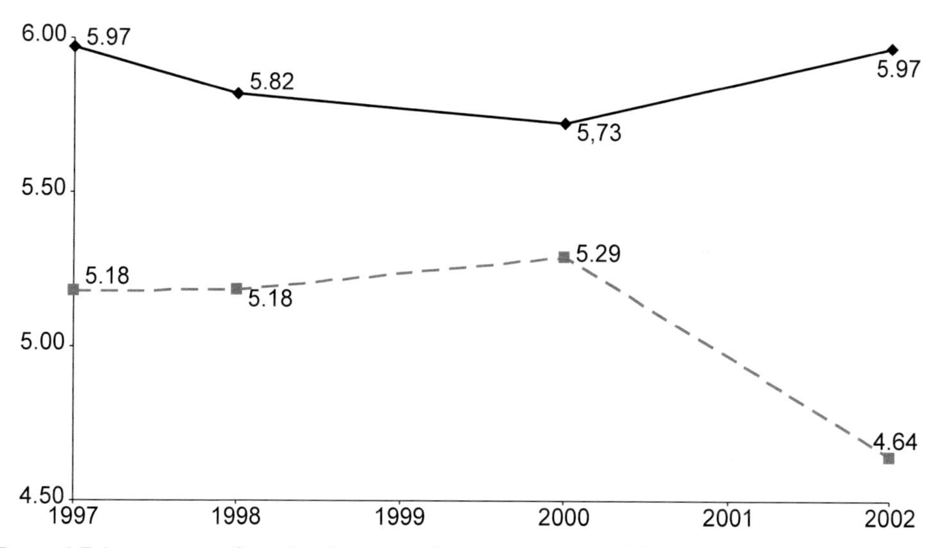

Figure 6.7. Importance of regular discussions between corporate R&D and end-users. Continuous line = R&D; dotted line = BUs/HQ.

The R&D staff should be going to the customer through Sales, not independently.

In 2002, when the competition was even more severe than the preceding years, the respondents were even more outright regarding this point:

R&D should not spend time and resources on finding out basic customer needs; this should be covered by the sales units and the BUs strategic functions.

And as a marketeer stated:

The corporate R&D center exists to expand our knowledge base and to assist/help in applying this knowledge. This puts high demands on the BUs to be active in defining customer needs well in advance and have an open mind as to how technology could be used to meet these needs. Instead of direct contact between the R&D staff and end-users, we need a strong connection between marketing and business development on the one hand and R&D on the other. They should meet on a monthly basis to discuss clear targets.

6.6 Concluding remarks

The results of the longitudinal study have been described in this chapter to answer research question two. The linear regression analysis revealed that the research variables R&D competencies and the R&D capabilities flexibility and communication correlate positively with external fit. This answers research question 2, while supporting the Proposition P(L)1 (see Section 4.4.2). Longitudinal analysis of the trends in answering patterns of R&D staff, as compared with BU customers, revealed that the initial gap between BU and R&D assessment disappears in the variables external fit, R&D competencies and responsiveness but not in the variables timeliness and communication. This indicates, that regular feedback led to a partial improvement in internal fit, thus answering Proposition P(L)2. The longitudinal analysis further revealed that one change introduced during the investigation, in particular contributed positively to strategic alignment, namely the change in R&D funding structure. This supports Proposition P(L)3 (see Section 4.4.2).

7. Discussion and conclusions

In this final chapter the main research question will be addressed.

How can technology-based firms achieve strategic alignment of innovation to business?

Looking at this question from the two main schools of strategic management, the industrial organization theory and the competence perspective, resulted in two possible sets of factors being identified that may be expected to affect strategic alignment: a firm's business environment, and a firm's own resources, competencies and capabilities. It was concluded in Chapter 2 that the following aspects have to be taken into consideration when studying the phenomenon of strategic alignment.

- Strategic alignment consists of the combination of internal and external fit.
- Strategic alignment can be studied along the dimension of the content of fit and the dimension of the process of creating alignment among (elements of) organizations.
- Strategic alignment is an inherently dynamic process, especially in the field of innovation.

To obtain complete insight into the phenomenon of the strategic alignment of innovation to business, it was decided to include these aspects in the empirical studies. To this end, it was decided to assess strategic alignment and its underlying factors with a combination of quantitative, objective and qualitative, subjective measurements.

Two studies were conducted to investigate the main research question empirically: a cross-industry study in ten multinational technology-based companies and a six-year longitudinal study within one technology-based company.

The cross-sectional survey design used for the cross-industry study had the advantage that differences among industries could be observed, but the disadvantage was that the dynamic aspect of alignment was beyond the scope of study. The survey was therefore well suited to investigating the content of fit but not the dynamic process of fit (see Venkatraman and Camillus, 1984, Section 2.7). This meant that we could study in-depth the formulation and implementation process of the innovation strategy at a certain moment in time. It should be noted, however, that the limited number of ten companies only allowed for tentative conclusions to be drawn.

The longitudinal study provided the possibility to follow the process of aligning innovation to business over time. The study design also enabled us to study all parts of the internal network that constitute the innovation function of the firm (R&D, BUs and headquarters) and was thus helpful in studying the process of aligning innovation to business from the perspective of the different partners. The high number of respondents and the relatively high response rate for an empirical management study may provide further confidence in the representativeness, and therefore, the generalizability of the findings.

In Sections 7.1 and 7.2 the results of the two studies are summarized, the research questions are answered and the empirical results are confronted with the propositions developed for each of the studies in Chapter 4. Section 7.3 provides the evaluation of the studies; the main research question is answered, the scientific and methodological contributions are indicated, and suggestions for further research are given. Section 7.4 elaborates on the implications of the studies for innovation management. Finally, Section 7.5 provides some concluding remarks.

7.1 The cross-industry study

7.1.1 The PGLC and the industry 'clockspeed'

The empirical results clearly confirm the validity of the PGLC as an indicator for the industry 'clockspeed'. Figure 5.3 shows a marked difference between short life cycle industries (SLCIs) and long life cycle industries (LLCIs). Companies in SLCIs that typically show time-spans between their subsequent product generations of up to five years generally report revenue contributions of new products of 70% or more, whereas companies in LLCIs, with PGLCs of seven years or longer, report significantly lower revenue contributions. More importantly, clear-cut differences in the proposed direction were found concerning the strategic alignment situation, innovation strategy, R&D competencies and the R&D capabilities between SLCIs and LLCIs. This clearly shows that the classification into SLCIs and LLCIs has a sound empirical basis. We therefore conclude that the PGLC proves to be a reliable indicator to make a relevant classification above the industry level. With regard to our first research question:

RQ1. What is the effect of the industry 'clockspeed' on the strategic alignment of innovation to business?

We can conclude that the market and technology forces in the business environment do indeed affect strategic alignment, and that their joint effect can be predicted when the length of the PGLC in a given industry is used as an indicator for the industry 'clockspeed'. In the following sections we look at these differences more in detail.

7.1.2 Strategic alignment

P(C)1. The level of strategic alignment between innovation and business is higher in SLCIs than in LLCIs, based on the closer market proximity in SLCIs.

The most important conclusion of the cross-industry study is that the strategic alignment between innovation and business is significantly better in SLCIs than in LLCIs, which can be explained by the closer market proximity in SLCIs.

Table 5.2 shows that all aspects of external fit are assessed significantly higher in SLCIs than in LLCIs. Companies in SLCIs, which encounter high market dynamism, clearly put more emphasis on monitoring market and technology trends than companies in the more stable LLCIs. For the same reason, it is understandable that market-technology road maps are used more intensively in SLCIs. The fact that the core technologies are less well defined in LLCIs and that the R&D portfolio is not based on a strategic technology and market vision is less easy to understand. Companies in LLCIs have to work with very complex systems and have to oversee relatively long time intervals. In pharmaceutical companies, a wrong strategic decision (for instance, a mistaken choice of therapeutic area) may cause years of R&D time to be conducted in vain. On the other hand, one may argue that especially at the beginning of such a long R&D process, like in pharmaceutical companies, there is more room for experimentation because of the low costs here compared to the preclinical phase and even more so compared to the clinical phase of the R&D process.

Table 5.3 shows that the results found on internal fit also support Proposition P(C)1. It shows that R&D management in SLCIs is more positive about their alignment to business strategy than in LLCIs. This is especially the case in their participation in the establishment of business plans and, although to a lesser extent, the involvement of the business units in R&D planning is more positively assessed in SLCIs. This could be expected because of the longer time interval between idea generation and market launch in LLCIs. It must be noted, however, that concerning the monitoring of key project parameters and the aligning of R&D objectives to business plans, the differences are small.

An interesting finding is that in all cases the respondents indicate that they think that the business units would, if asked, assess the level of internal fit lower than they do themselves (see Table 5.4). LLCI respondents in particular thought that there was friction in the relationship between R&D and business units from the BU point of view. It must be frustrating for R&D management in LLCIs to perceive so little positive feedback on their attempts to align themselves to business. One of the reasons for this might be that the business units in SLCIs typically use technologies that can be inserted into systems, at least as engineering prototypes, within one or two years, whereas the immature technologies in LLCIs may require three to ten years of additional development before they can be used by the business units. Also the finding that corporate support to reinforce the R&D interests of the business units was absent in most of the participating companies is remarkable.

If we compare the results on alignment with those found in a survey conducted by Cooper (1999) among 203 high- and medium-tech firms, we can conclude that self-assessment by the participating companies in our study is significantly higher than the average found in Cooper's study. The technology leadership status of the participating companies can be shown by the fact that most of them score in or just below the top 20% of Cooper's research population.

7.1.3 R&D strategy

P(C)2. The relatively high market dynamism in SLCIs will lead to a more exploitation-oriented R&D strategy. The relatively high technology complexity in LLCIs will lead to a more exploration-oriented R&D strategy.

Proposition P(C)2 is supported by the quantitative data on the sales revenue of new products, the percentage of the total R&D budget spent on research and the educational level of the R&D staff (see Table 5.5) but not by qualitative self-assessment (see Table 5.6). In the qualitative data on this aspect, we find a somewhat more radical orientation in LLCIs but it is still far from significant. In this respect, it is interesting to note that R&D managers in LLCIs report having major problems in getting funding for research and technology development, significantly more so than those in SLCIs.

These findings should perhaps be interpreted as an indication of a shift to more open forms of innovation because a number of companies in the industries in which the participating companies were active at the time of investigation decided to stop basic technology development, buying instead knowledge and capabilities from specialized research firms, and knowledge institutions, especially large RTIs (Research and Technology Institutes). Most of the companies in the present study, however, made the strategic decision to stay in research and technology development because of the threat of losing absorptive capacity, weakening the technology knowledge base of the company and losing negotiating power vis-à-vis research firms and RTIs.

Table 5.6 shows another unexpected finding, especially for SLCIs, namely that speed in conducting R&D projects is not emphasized over budget. Of course, it is always wise to stay within budget but in the light of the vital importance of speed to stay ahead of the competition in SLCIs, and the comparatively low cost of R&D compared to being late on the market, this emphasis on budget is remarkable. Beswick (cited in Twiss, 1992) concludes, based on figures from the electronics industry, that exceeding the R&D budget by 50% (which normally leads to critical discussions with corporate headquarters) may lead to a decrease in profits of only 3.5%, whereas being six months late on the market may cost 33% of profits, if measured over the whole life-span of the product generation.

Table 5.7 shows that in accordance with Proposition P(C)2, the attention paid to value engineering in the participating companies is significantly larger in SLCIs. The participating companies indicated having achieved about a 15% saving in their present products and processes as a result of changes adopted over the last three years. That many companies emphasize the importance of value engineering was also shown in the EU ESPRIT project of 1996, where 44% of the companies indicated paying attention to value engineering. In the automotive and chemical industries these figures are even higher, 55% and 67% respectively. The level of attention paid to early development by senior management is used as a proxy for the emphasis that senior management places on portfolio planning. Here the proposition is tentatively supported.

The finding concerning the number of patents per R&D investment is in clear contrast to Proposition P(C)2. The number of patents is about six times higher in SLCIs than in LLCIs. This unexpectedly high number of patents per R&D investment may reflect the essential position that industrial property rights has gradually attained in SLCIs. As indicated in the structured interviews, more and more patents are filed to defend core technologies, hinder competitor activities by patenting around a competitor's patent, but also for cross-licensing trade with other companies. Interestingly, the cross-industry companies did not seem very keen on getting the best return on their R&D investments. Only two companies indicated that they had a strategy on this. These two companies recoup 5-10% and 15-20% of their R&D budget in royalties for the use of their patents.

7.1.4 R&D competencies

P(C)3. The more exploitation-oriented R&D strategy in SLCIs will lead to R&D competencies being more focused on in-house knowledge. The more exploration-oriented R&D strategy in LLCIs will lead to R&D competencies being more focused on open innovation through collaboration with external knowledge sources.

Table 5.8 shows that, in accordance with Proposition P(C)3, SLCIs put more emphasis on in-house competency planning. In particular, the cross-functionality of competency monitoring is assessed significantly higher in SLCIs than in LLCIs. Some respondents indicated in the structured interviews that they found competency listing to be especially useful for the recognition of gaps in. the core competencies that they needed to acquire in order to achieve success in the market. Table 5.9 shows that in contrast to Proposition P(C)3, not much difference was found between SLCIs and LLCIs regarding the level of cooperation with suppliers and customers. But in accordance with Proposition P(C)3, LLCIs made significantly more use of specialized R&D contractors such as research firms and knowledge institutions.

7.1.5 R&D capabilities

P(C)4. The more exploitation-oriented R&D strategy in SLCIs will lead to R&D capabilities being more directed at increasing customer orientation and speed to market. The more exploration-oriented R&D strategy in LLCIs will lead to R&D capabilities being more directed towards external knowledge acquisition.

Figure 5.4 shows that the cycle time reduction realized in the preceding three years in LLCI companies was considerably higher than in SLCI companies. This was an unexpected finding and in clear contrast with Proposition P(C)4, seen in the light of the vital importance of speed in SLCIs. That these differences were not mere coincidence is indicated by the fact that the cycle time reduction found in the present study was in line with that found in a much larger group of 200 strategic business units in SLCIs (McGrath, 1996). Perhaps this unexpected result can be explained as catching up of the R&D centers in LLCIs with the R&D centers in SLCIs that may already have achieved major improvements in cycle time reduction in the preceding period. It takes considerable effort to reduce R&D cycle time further when the first major steps have already been taken.

It was also expected that R&D centers in SLCIs would be more responsive to BU requests because of the higher market dynamism they are confronted with but in fact no difference was found in the answers between the SLCI and LLCI R&D labs.

Table 5.10 shows that, in accordance with Proposition P(C)4, the level of internal communication with BUs, manufacturing and marketing and sales is assessed consistently higher in SLCIs than in LLCIs. In the structured interviews it was indicated that LLCIs have to overcome the physical and cultural distances between R&D and manufacturing. A number of LLCI companies had therefore set up special technology transfer meetings between R&D and manufacturing and included manufacturing engineers in their R&D project teams. One of the companies has solved this problem by organizing additional assignments for senior R&D managers as principal interfaces between R&D and manufacturing; they spend part of their time at the R&D lab and the rest at the manufacturing plant(s). Table 5.10 also shows that in many of the R&D centers there is the feeling that cross-functional communication with marketing should improve. The communication gap between R&D and marketing is not new. This is especially true in LLCIs where the conception of an innovative idea and final market launch are typically ten or more years apart (se Chapter 1, the R&D marketing gap).

Table 5.11 shows that, in accordance with Proposition P(C)4, the level of customer orientation is higher in SLCIs than in LLCIs. In the SLCI Centers project ideas are consistently evaluated in terms of their value to customers. Also in the structured interviews and from the answers to Questionnaire I it was apparent that SLCI Centers put more emphasis on direct R&D contact with the external customer; on average 5% to 10% of working time was spent on direct contact. Finally, in accordance with Proposition P(C)4 about more scientific communication in LLCIs, 80% of the R&D centers in LLCIs actively encouraged scientific publications and presentations at international conferences.

7.2 The longitudinal study

7.2.1 R&D competencies and capabilities

P(L)1. A higher level of internal fit, in terms of the perceived adequacy of the R&D competencies and capabilities, will be positively related to a higher level of external fit, in terms of R&D alignment to market and technology needs.

The linear regression analysis (Table 6.5) clearly indicates that 45% of the total variance of strategic alignment (external fit) can be explained by the competency and capability variables and a number of control variables. It shows that the R&D competencies (carrying out the correct research and doing so in the right way) in the six-year period of the investigation became much more aligned to BU needs. From the R&D capabilities, the timeliness of R&D projects comes out as the most important. This is not surprising if we consider the marked decrease in project cycle time that was achieved by LLCI companies in general (Figure 5.4). Table 6.6, and Figures 6.2 to 6.4 show that the assessment of R&D competencies and the R&D capabilities responsiveness and timeliness, all show a similar trend over time, namely an increase in BU appreciation and/or a significant reduction in the initial gap between BU/HQ assessment and R&D staff self-assessment.

It must be noted, however, that the R&D capability R&D-BU communication showed no improvement over time (Table 6.6 and Figure 6.5). The BU assessment remains more or less stable at a level of 5, while R&D staff self-assessment was declined steadily over the years. However, all the respondents indicated that communication between R&D and business was important, and this figure gradually increased to more than six on a seven-point scale in 2002 for both the R&D and the BU/HQ respondents (see Figure 6.6), We therefore presume that this finding should not be interpreted as a signal that information exchange is not relevant, but rather as an indication that 'good is good enough' in the eyes of the BUs/HQ.

Figure 6.7 shows that the BU/HQ assessment of the importance of R&D staff directly communicating with end-users clearly decreased over the investigation period. The BUs are clearly not in favor of the idea of R&D staff having regular contact with end-users independent of the Business Units. Their main fear is that R&D staff will offer solutions to end-users before a commercial price can be negotiated. With regard to our second research question:

RQ2. How can strategic alignment of innovation to business be achieved and maintained over time?

We can conclude that R&D competencies and the R&D capabilities of flexibility and communication have a clear impact on strategic alignment.

7.2.2 Strategic alignment through structured feedback

P(L)2. Structured feedback on R&D competencies and capabilities will help to achieve and maintain internal fit.

What was most important was the fact that the BU assessment concerning the strategic alignment of R&D projects improved considerably over time (Figure 6.1). While there was a considerable gap in 1997 between BU assessment and R&D staff self-assessment, this gap had totally disappeared in 2002. In 1997 the BU/HQ assessment of the strategic alignment of

R&D projects to business was quite negative. This was not unexpected because, as was already stated in Chapter 1, there often appears to be a certain tension between R&D and business units, the long-term orientation needed for exploration and the uncertainty of its outcome being at odds with the predictability needed for executing day-to-day activities efficiently (Roberts, 1995; Glass *et al.*, 2003; Park and Gil, 2006). Respondents from headquarters and BU directors were especially critical in 1997. The respondents from headquarters and the BUs predominantly perceived R&D as a drain on company resources, and they were the ones having to pay the bill. In clear contrast to this, in 2002 the BU/HQ assessment was even a bit higher than the corporate R&D staff's self-assessment, and within the BU/HQ segment it was top management that showed the most positive assessment. Concerning the expected link between improved strategic alignment and business performance, it is worth noting that sales figures of the company under study measured two years after the respective surveys show a rise of 22% from 1999 through 2004, although such figures have to be interpreted with some caution for the obvious reason that they are influenced by many other factors than R&D to business alignment alone.

From the 'Communication web' stream of research (see Section 4.4.2) it was expected that communication would come out as one of the most prominent factors to affect alignment. At first glance, the results presented in the previous section do not support this hypothesis. The reader should bear in mind, however, that in the present study in fact two forms of communication were investigated, the structured feedback on the quality of the R&D center's competencies and capabilities, and one of these capabilities, being the research variable of R&D communication, which in essence covered basic forms of project progress and outcome reporting. Our findings clearly indicate that the communication in the form of the structured feedback acted as an effective tool to enhance alignment. From our findings it may therefore be concluded, that communication to be effective, should go beyond the regular project progress reports, and should include feedback on the relevant competencies and capabilities of the corporate R&D center.

7.2.3 Management methods for improving strategic alignment

P(L)3. An R&D funding structure that effectively balances the exploitation and exploration function will help to achieve and maintain internal fit.

Figure 6.1 clearly shows that, although the introduction of the Balanced R&D Score Card and the system of market-technology road mapping certainly had a positive impact on strategic alignment, it was clearly the change in governance structure in 1999 that brought about the change in R&D funding structure that had the largest positive impact on the strategic alignment situation. This shifted the locus of control over the R&D portfolio from R&D exclusively to a joint responsibility of R&D and the BUs and headquarters. BU funding effectively transforms the business units into external customers for corporate R&D, leading to improved alignment of R&D and business objectives. In management literature there has been a fierce debate as to whether a shift from corporate to business unit funding would not destroy the long-term R&D orientation of these companies. It is concluded that a system that effectively balances short-term orientation (via BU unit-funding) and long-term orientation (via Technology Board-funding in which R&D management, top management of headquarters, and BU directors jointly decide on long-term R&D projects) is effective in providing strategic alignment between R&D and business.

7.3 General conclusion

Combining the findings of the cross-industry and the longitudinal study enables us to provide an answer to our main research question.

How can technology-based firms achieve strategic alignment of their innovation processes with their business activities?

The answer has to be that firms must realize that the alignment of innovation to business is strongly influenced by the combined effects of the market and technology forces that are at work in the industries in which they operate, as reflected by the product generation life cycle (PGLC). But they should rely on their own strengths to improve their level of strategic alignment by providing regular feedback on their R&D competencies and capabilities by carefully balancing the influence over the R&D portfolio of the business units and corporate headquarters on the one hand, and R&D on the other, and by implementing the management tools such as those identified in the present studies.

7.3.1 Theoretical contributions

The present study clearly shows the importance of integrating the industrial organization and the competence perspectives to gain a more comprehensive understanding of the phenomenon of strategic alignment, which is one of the core constructs used in the field of strategic management. While the present study is limited to the alignment of innovation to business strategy, we presume that a similar approach could be successfully applied to study alignment in other fields of strategic management.

Our study adds to each of the two perspectives used: to the competence perspective by defining the concepts of competencies and capabilities at the sub-firm level of the innovation function, and to the industrial organization theory by introducing the product generation life cycle (PGLC) as a relevant indicator to classify companies. By proposing the PGLC as a new theoretical concept, the present study creates new opportunities to analyze and compare a firm's conduct at a higher aggregation level than the specific industry.

The contribution of this study is to the competence perspective can be considered a reply to Foss (2005), who criticized the competence perspective for focusing too heavily on the company level by turning to organizational capabilities as a key construct (e.g. Eisenhardt and Martin, 2000; Winter, 2003). He argues, that sustained competitive advantage, a firm-level phenomenon, is now directly explained in terms of firm-level competencies and capabilities, ignoring the lower than firm-level competencies and capabilities. By defining competencies and capabilities at the sub-firm level of the innovation function, which is an approach advocated by Coombs (1996), the present study overcomes this problem. The results of both empirical studies clearly demonstrate that this is a meaningful way forward in improving our understanding of a firm's strategic conduct from a lower level of analysis.

By including a longitudinal design, the study answers requests by researchers for a longitudinal investigation of strategic alignment (e.g. Venkatraman, 1989; Miller, 1992; Coombs, 1996; Zajac *et al.*, 2000) and for studies that investigate the evolution of dynamic capabilities over time (Helfat and Peteraf, 2003).

7.3.2 Methodological contributions

The combination of a cross-sectional study with a longitudinal study had a number of advantages. First, it provided an optimal triangulation since the weaknesses in one study (only one company in the longitudinal study; only one point in time in the cross-industry study) were compensated for by the strengths of the other. Perhaps even more important is the fact that where the cross-industry study enabled us to evaluate the proposed sign of the relationships and the relative strengths of the different independent variables, the longitudinal study could inform us about causal relationships using its 'quasi-experimental' design character.

It can also be concluded that the transversal cross-section of the best companies in different industries proved to be a successful way of shedding light on the complex world of the strategic alignment of innovation to business, and that the choice of different angles of analysis, which each provided their essential contribution to the elucidation of the underlying problem, adds to the reliability of the findings.

The combination of 'subjective' judgments about the quality of the strategic alignment situation, and the R&D management systems with more 'objective' measures of performance and the management and communication systems that were actually in place, has proved to be a very successful method of investigation. The subjective measures offer the unique opportunity to gain an inside view of how R&D staff and their customers in the business units perceive the alignment situation and the achieved level of R&D competencies and capabilities, while the objective measures provided a check on the reliability of the answers.

7.4 Implications for innovation management

In this section we provide suggestions for innovation management based on the theoretical and practical insights presented in the preceding sections. These can assist managers to answer the following questions:

What are the main factors that CTOs and R&D managers have to take into consideration if a technology-based prospector company wants to align its innovation strategy to its business? What tools do they need to maintain strategic alignment once it is established, and how can eventual pitfalls be avoided?

First, it is advised that CTOs and R&D managers take special notice of the industry clock speed of the industry sector(s) in which they operate, by looking at the length of the product generation life cycle (PGLC) of their main products.

This is important, because the present study shows, that not only the character of a company's innovation strategy, but also the achieved level of strategic alignment of innovation to business is influenced by the length of the PGLCs of its main products. Our data indicate that companies operating in industries characterized by relatively short product generation life cycles (SLCIs, shorter than 6 years) have problems in maintaining the long term technology-knowledge base of their firms, whereas companies operating in industries characterized by relatively long product generation life cycles (LLCIs, longer than 6 years) tend to have problems in aligning their innovation strategy to the short term needs of their business units. Knowing this, it might be wise, to counterbalance these tendencies by putting special efforts on maintaining the technology-knowledge base in SLCIs and putting special emphasis on cycle time reduction and the alignment of innovation to business strategy in LLCI companies. CTOs and R&D management can use the 'PGLC-dependent' relationship of revenue contribution of new products introduced to the market in the preceding three years to benchmark their results with the first class technology-based companies included in the present study (see Figure 5.3).

Second, the two main tools identified in the present study available to companies to achieve strategic alignment, and, more importantly, maintain it over time are an R&d funding structure that effectively balances the exploitation and exploration function, and regular structural feedback on the perceived adequacy of the R&D competencies and capabilities.

The data from the present study show, that BU funding effectively transforms the business units into external customers for corporate R&D, leading to improved alignment of R&D and business objectives. Combining this with technology board funding helps to maintain the balance between the short term business needs of the Bus and the long term technology needs of the firm. Introducing technology board funding can therefore be especially important for SLCI companies to safeguard their long-term technology outlook.

Our study further clearly demonstrates, that for maintaining alignment between innovation and business, structural feedback from the Bus and headquarters to corporate R&D is essential. Upon this information CTOs and R&D management can base targeted management methods (see below).

Third, the present study identified a number of tools that can be utilized to improve strategic alignment

- Market-technology roadmaps and Balanced R&D Score Cards.
- Staff exchange between R&D and business.
- R&D project evaluation in terms of their alignment to business.
- R&D project prioritization by BUs as to their value to customer s.
- Cross-functional participation in R&D planning.
- R&D project parameters discussed with BUs;
- R&D participation in BU business plan formulation.

The findings of the longitudinal study (Chapter 6) indicate that the Introduction of Market-Technology Road Maps and Balanced R&D Score Cards significantly contributed to achieve strategic alignment. Staff exchange came out as an important tool to improve the communication between corporate R&D and the business units. The other above mentioned management methods were all identified in the cross-industry study as methods to improve strategic alignment. Especially LLCI companies can learn from SLCI companies who already use these methods to align R&D to the short term business needs.

Finally, all efforts to achieve and maintain strategic alignment between innovation and business have to be embedded in the overall process of strategy making.

As we have seen in the Chapters 2 and 3 this process consists of a cycle of strategic intent, strategic choice, strategy implementation and strategy evaluation.

In the strategic choice phase, strategic alignment between innovation and business is at stake when general and R&D management take the often difficult decisions about which R&D projects receive resources and which do not. Insights into the firm's innovative potential and the barriers to innovation are necessary to make these choices effectively. Special methods have to be used to gain an idea as to how much money should be spent on R&D, e.g. as compared to main competitors, and the amount of R&D budget to be spent on different innovation areas, e.g. the percentage of sales based on new products introduced on to the market in the last three years and the percentage of the R&D budget that stems from corporate funding as compared to business unit and external funding (e.g. EU subsidies, innovation carried out for other companies, and licensing royalties). Also policy issues have to be addressed e.g. concerning know how, patents, and trade secrecy.

When in the strategy implementation phase the firm's innovation strategies are put into action, the way in which questions, as to how to ease technology transfer, for example by moving personnel between functional areas (e.g. through cross-functionality of project teams including R&D, marketing, manufacturing, purchasing, law and environmental experts) or between mainstream activities and new ventures (spin-off policy), are answered will have an impact on the achieved level of strategic alignment. Management tools, such as the R&D Balanced Score Card, and the open innovation matrix (Table 7.1) can serve as comprehensive management tools for supporting these decisions.

The first step in framing the open innovation matrix is to establish the competitive impact of the company's technologies by dividing them into emerging (embryonic) technologies, pacing technologies (with an expected short- or medium-term competitive impact), key (core competence) technologies, and base (widespread and shared) technologies (Roussel *et al.*, 1991), To this end, for each aspect of technology the following questions have to be answered.

- Does this technology have the potential for competitive differentiation?
- Could it become critical to the firm?
- And what is its market value?

The second step is to assess the in-house capabilities of each of the technologies using the categories weak, moderate and strong. By combining the competitive technological impact with the in-house technological capabilities we get the open innovation matrix presented in Table 7.1.

An emerging technology may have a competitive impact in the future. If a technological capability is strong, it has to be optimized reinforcing its potential competitive advantage. If the internal technological capability is moderate or weak, catching up may be necessary. However, uncertainty demands for scanning the R&D environment, i.e. having many partners and flexible relationships, preferably in strategic partnerships and alliances, or via contract

Table 7.1. Open innovation matrix.

Competitive impact	**In-house technological capabilities**		
	Weak	**Moderate**	**Strong**
Emerging technology	scan	scan/collaborate	collaborate
Pacing technology	collaborate	share risks	in-house
Key technology	optimize	optimize	in-house
Base technology	outsource	outsource/exchange	sell/exchange

Adapted from Omta and Folstar (2005).

research and sponsoring knowledge institutions. In all cases, adequate patent protection strategies need to be considered. A pacing technology may have strong competitive impact in the short or medium term. If a firm's technological capability is relatively strong, the bias should be towards doing work in-house. Extra investments might be required for research into the application of the technology in new products and markets. If a firm's technological capability is moderate, sharing the risk by strategic alliances with partner firms makes the most sense. If one's technological capability is weak, acquiring licenses or joint development may be viable alternatives. Pacing technologies need the utmost management care, especially if the technology is maturing rapidly because it might become essential for tomorrow's business. It is therefore necessary to scan research efforts of competitors and potential technology sources intensively. Technologies need to be carefully protected. Generally speaking, a company should own key technologies that are critical to current competitiveness. If a firm's technological capability is weak or only moderate in a key technology area, it should get extra technological capability for building in-house R&D strengths by acquiring or introducing a substitute technology. For non-critical base technologies, outsourcing might be the most appropriate choice if one's technological capability in the field is weak. If it is moderate, it may serve as a means of exchange in a partnership. And if it is strong, it may either serve as a means of exchange or be sold to focus the internal technological capabilities on key technologies. It is important to remember that the open innovation matrix does not offer a static model. For instance, by acquiring extra technological capacity in a field of key technology where one's technological capability is weak, one should gradually shift from the left-hand side to the right-hand side of the third row in the matrix.

In the evaluation phase, the performance of the different innovations is assessed to see how well the innovation strategy has been implemented and whether the firm's objectives have been met. Strategy evaluation should be conducted on a continuous basis, not only to review the whole R&D portfolio but also to monitor the levels of internal and external fit as perceived within the corporate R&D-HQ-BU triangle.

Feedback on levels of internal and external fit as well as on perceived levels of R&D competencies and capabilities should be made available to corporate R&D and its internal business unit customers. Top management involvement is paramount to constantly monitor the 'fit' between the firm's innovation strategy, assets and resources and the rapidly changing business environment, markets and technologies.

7.5 Concluding remarks

Numerous researchers have stressed the importance of aligning innovation to business strategy (e.g. Burns and Stalker, 1961; Lawrence and Lorsch, 1967; Ginsburg and Venkatraman, 1985; Miles and Snow, 1994; Verdú Jover *et al.*, 2005). The importance of a flexible organization to proactively react to changing environmental situations at strategic, tactic and operational level (e.g. Volberda, 1992) and the importance of maintaining an open and extensive R&D network are also stressed in many studies (e.g. Biemans, 1992; Della Valle and Gambardella, 1993; Albertini and Butler, 1994). However, until now only limited evidence has been presented derived from empirical studies of the real world of management practice as to how such alignment can be achieved and maintained over time.. It is the merit of the present study that it fills up this gap by combining a cross-sectional and a longitudinal approach to provide an empirically based comprehensive understanding of the phenomenon of strategic alignment of innovation to business. Companies must forge an innovation strategy that's aligned with its overall strategy, choose the projects with the best value propositions, manage the system efficiently so it doesn't waste time or resources, and commercialize innovations well, with everyone working together as a team (Jaruzelski *et al.*, 2005).

We sincerely hope that the theoretical insights and practical management tools and methods described in this book will help corporate and innovation managers to accomplish the important task of aligning innovation to business.

References

Aaker, D., 1995. Strategic Market Management. John Wiley & Sons, New York etc.
Aaker, D. and B. Mascarenhas, 1984. The Need for Strategic Flexibility. Journal of Business Strategy, 5 (2), 74-82.
Ahuja, G., 2000. The Duality of Collaboration: Inducements and Opportunities in the Formation of Interfirm Linkages. Strategic Management Journal, 21 (3), 317-343.
Albertini, S. and J. Butler, 1994. The R&D Networking Process in a Pharmaceutical Company. The R&D Management Conference, Manchester.
Albright, R. and T. Kappel, 2003. Roadmapping in the Corporation. ResearchTechnology Management, 46 (2), 31-40.
Alchian, A.A., 1950. Uncertainty, Evolution and Economic Theory. Journal of Political Economy, 58 (3).
Aldrich, H.E., 1979. Organizations and Environments. Prentice-Hall, Englewood Cliffs, New Jersey.
Allen, T.J., 1977. Managing the Flow of Technology: Technology Transfer and the Dissemination of Technological Information within the R&D Organization. MIT Press, Cambridge, Mass.
Amabile, T.M., 1996. Creativity in Context. Boulder, CO: Westview Press.
Anderson, P. and M.L. Tushman, 1990. Technological Discontinuities and Dominant Designs: A Cyclical Model of Technological Change. Administrative Science Quarterly, 35(4), 604-33.
Andrews, K.R., 1971. The Concept of Corporate Strategy. Irwin, Homewood Il.
Angle, H.L., 1989, Psychology and Organizational Innovation. In: A.H. Van de Ven, H.L. Angle and M.S. Poole (Eds.), Research on the Management of Innovation: The Minnesota studies. Harper and Row, New York, 135-170.
Ansoff, H.I., 1965. Corporate Strategy: An Analytical Approach to Business Policy for Growth and Expansion. McGraw-Hill, New York.
Ansoff, H.I., 1982, Societal Strategy for the Business Firm. In: H.I. Ansoff, A. Bosman and P.M. Storm (Eds.), Understanding and Managing Strategic Change, New York.
Arthur D. Little, 1991. The Arthur D. Little Survey on the Product Innovation Process. Arthur D. Little, Cambridge Mass.
Babbie, E.R., 2003. The Practice of Social Research. Wadsworth, Publ., Belmont.
Bain, J.S., 1954, Conditions of Entry and the Emergence of Monopoly. In: E.H. Chamberlain (Ed.), Monopoly and Competition and their Regulation. Macmillan, London, 215-244.
Bain, J.S., 1956. Barriers to New Competition. Harvard University Press, Cambridge Mass.
Barney, J., 1991. Firm Resources and Sustained Competitive Advantage. Journal of Management, 17 (1), 99-120.
Baron, R.M. and D.A. Kenny, 1986. The Moderator-Mediator Variable Distinction in Social Psychological Research: Conceptual, Strategic, and Statistical Considerations. Journal of Personality and Social psychology 51, 1173-1182.
Bart, C.K., 1991. Controlling New Products in Large Diversified Firms: A Presidential Perspective. Journal of Product Innovation Management, 8 (1), 4-17.
Bayus, B.L., 1994, Are Product Life Cycles Really Getting Shorter? Journal of Product Innovation Management 11, pp. 300-308.

Biemans, W.G., 1992. Managing Innovation within Networks. Routledge, London, New York.

Bonner, J.M., R.W. Ruekert and O.C. Walker Jr, 2002. Upper Management Control of New Product Development Projects and Project Performance. The Journal of Product Innovation Management, 19, 233-245.

Booz-Allen and Hamilton, 1982. New Products Management for the 1980s. Booz- Allen and Hamilton, New York.

Brockhoff, K., A.K. Chakrabarti, and J. Hauschildt (Eds.), 1999. The Dynamics of Innovation. Springer, Heidelberg.

Brown, S.L. and K.M. Eisenhardt, 1995. Product Development: Past Research, Present Findings and Future Directions. Academy of Management Review, 20 (2), 343-378.

Brown, S.L. and K.M. Eisenhardt, 1998. Competing on the Edge. Strategy as Structured Chaos. Harvard Business School Press, Boston, Mass.

Buderi, R., 2000. Funding Central Research. Research.Technology Management, 43 (4), 18-25.

Burgelman, R.A. and R. Rosenbloom (Eds.), 1997. Research on Technological Innovation. Management and Policy, 6. JAI Press, Greenwich.

Burgelman, R.A., 1988. Strategy-Making as a Social Learning Process: The Case of Internal Corporate Venturing. Interfaces, 18 (3), 74-85.

Burns, T. and Stalker, G.M., 1961. The Management of Innovation. Tavistock, London.

Burton, R.M. and B. Obel, 1998. Strategic Organizational Diagnosis and Design: Developing Theory for Application, Kluwer Ac. Publ., Dordrecht.

Calantone, R.J., S.T. Cavusgil and Y. Zhao, 2002. Learning Orientation, Firm Innovation Capability, and Firm Performance. Industrial Marketing Management, 31, 515-524.

Caloghirou, Y., Kastelli and A. Tsakanikas, 2004. Internal Capabilities and External Knowledge Sources: Complements or Substitutes for Innovative Performance? Technovation, 24, 29-39.

Capron, L. and W. Mitchell, 1998. Bilateral Resource Redeployment and Capabilities Improvement Following Horizontal Acquisitions. Industrial and Corporate Change, 7 (3), 453–484.

Carr, N.G., 1999, Forethought: Visualizing Innovation. Harvard Business Review 77(5), 16.

Chakravarthy, B.S., 1982. Adaptation: A Promising Metaphor for Strategic Management. Academy of Management Review, 7, 35-44.

Chandler, A.D., 1962. Strategy and Structure: Chapters in the History of the Industrial Enterprise. MIT Press, Cambridge Mass.

Chesbrough, H.W., 2003. Open Innovation: The New Imperative for Creating and Profiting from Technology. Harvard Business School Press, Cambridge Mass.

Chester, A.N., 1994. Aligning Technology with Business Strategy. Research. Technology. Management, January-February, 25-32.

Christensen, C., 1997. The Innnovator's Dilemma. Harvard Business School Press, Boston, Mass.

Christensen, H.K. and C. Montgomery, 1981. Corporate Economic Performance: Diversification Strategy versus Market Structure. Strategic Management Journal, 2, 327-343.

Clark, K.B. and T. Fujimoto, 1991. Product Development Performance. Harvard Business School Press, Boston Mass.

Clark, M.A., 2006. Mastering the Innovation Challenge, Booz-Allen and Hamilton, Tysons Corner, Virginia.

Cobbenhagen, J., 1999. Managing Innovation at the Company Level. PhD Thesis University of Maastricht, The Netherlands.

Cohen, W. M. and D.A. Levinthal, 1990. Absorbtive Capacity: A New Perspective on Learning and Innovation. Administrative Science Quarterly, 35, 128-152.

Collis, D.J., 1991. A Resource Based Analysis of Global Competition: The Case of the Bearings Industry. Strategic Management Journal, 12, 49-68.

Collis, D.J. and D. Montgomery, 1995. Competing on Resources: Strategy in the 1990s. Harvard Business Review, 73 (4), 118-129.

Conner, K., 1991. A Historical Comparison of Resource-Based Theory and Five Schools of Thought within Industrial-organization Economics: Do We Have a New Theory of the Firm? Journal of Management, 17 (1), 121 – 154.

Coombs, R., 1996. Core Competencies and the Strategic Management of R&D. R&D Management 26 (4), 345-355.

Cooper, D.R. and C.W. Emory, 1995, Business Research Methods. Irwin, Homewood, Ill.

Cooper, R., 1999. The Invisible Success Factors in Product Innovation. Journal of Product Innovation Management, 16, 115-133.

Cooper, R., S. Edgett and E.J. Kleinschmidt, 2001. Portfolio Management for New Product Development: Results of an Industry Practices Study. R&D Management, 31 (4), 361-380.

Cooper, R.G. and E.J. Kleinschmidt, 1995. Performance Typologies of New Product Projects. Industrial Marketing Management, 24, 439-456.

Cox, W., 1967, Product Life Cycles as Marketing Models. Journal of Business 40 (Oct.), pp. 375-384.

Cordero, R., 1990. The Measurement of Innovation Performance in the Firm: An Overview. Research Policy, 19, 185-192.

Coyn, K.P., 1985. Sustainable Competitive Advantage - What is it, What it isn't. Business Horizons, 29 (1), 54-61.

Damanpour, F. and W.M. Evan, 1990. The Adoption of Innovations over Time: Structural Characteristics and Performance of Organizations, National Decision Science Institute Conference, San Diego, Cal.

Davenport, T.H. and V. Grover, 2001. General Perspectives on Knowledge Management: Fostering a Research Agenda. Journal of Management Information Systems, Summer 2001, 5-21.

Davis, J.T., 1997. Top Companies. The Forbes Annual Review of Today's Leading Businesses, John Wiley & Sons, New York, etc.

Day, G.S. and P. Nedungati, 1994. Managerial Representations of Competitive Advantage. Journal of Marketing, 58 (2), 31 – 44.

Della Valle, F. and A. Gambardella, 1993. Biological Revolution and Strategies for Innovation in Pharmaceutical Companies, in R&D Management, 23 (4).

Dess, G.G. and Beard, D.W., 1984. Dimensions of Organizational Task Environments. Administrative Science Quarterly 29 (1), 52-73.

Dierickx, I. and K. Cool, 1989. Asset Stock Accumulation and Sustainability of Competitive Advantage. Management Science, 35 (12), 1504-1511.

Donnellon, A., 1993. Crossfunctional Teams in Product Development. Journal of Product Innovation Management, 10, 377-392.

Dosi, G., 1988. Sources, Procedures and Microeconomic Effects of Innovation. J. Econ. Lit., 26, 1120-1171.

Dosi, G., Freeman, C., Nelson, R.R., Silverberg, G. and Soete, L. (Eds.), 1988. Technical Change and Economic Theory. Pinter, London.

Dougherty, D., 1992. Interpretive Barriers to Successful Product Innovation in Large Firms. Organization Science, 3 (2), 179-202.

Drazin, R. and A.H. Van de Ven, 1985. An Examination of Alternative Forms of Fit in Contingency Theory. Administrative Science Quarterly, 30, 514-539.

Drucker, P.F., 1985. Innovation and Entrepreneurship: Practices and Principles. Harper & Row, New York.

Duysters, G. and C. Lemmens, 2003. Alliance Group Formation: Enabling and Constraining Effects of Embeddedness and Social Capital in Strategic Technology Alliance Networks. International Studies of Management and Organization, 33 (2), 49 - 68.

Edler, J., F. Meyer-Krahmer and G. Reger, 2002. Changes in the Strategic Management of Technology: Results of a Global Benchmarking Study. R&D Management, 32 (2), 149-164.

Eisenhardt, K.M., 1989. Building Theories from Case Study Research. Academy of Management Review. Vol. 14, No. 4, pp 532-550.

Eisenhardt, K.M. and J.A. Martin, 2000. Dynamic capabilities: What are they? Strategic Management Journal, 21 (10-11), 1105-1121.

Ernst, H., 2002. Success Factors of New Product Development: A Review of the Empirical Literature. International Journal of Management Reviews, 4 (1), 1-40.

Evered R., 1983. So What is Strategy? Long Range Planning, 16 (3), 57-72.

Fagerberg, J., D.C. Mowery and R.R. Nelson (Eds.), 2004. The Oxford Handbook of Innovation. Oxford University Press, Oxford.

Fine, C.H., 1998. Clockspeed. Winning Industry Control in the Age of Temporary Advantage. Perseus Books Group, New York.

Fisher, J. and R. Pry, 1971. A Simple Substitution Model of Technological Change. Technological Forecasting and Social Change 3, 75-88.

Fortuin, F.T.J.M., F.H.A. Janszen and S.W.F. Omta, 2005. The CUSVALIN Model. A Longitudinal Study of Customer Value Learning in Innovation, in Technology Management: A Unifying Discipline for Melting the Boundaries. IEEE, New Jersey [IEEE Cat. Number 05CH37666, ISBN 1-890843-11-3], 253-262.

Fortuin, F.T.J.M. and S.W.F. Omta, 2006. Aligning R&D to Business Strategy. A Longitudinal Study from 1997 to 2002. In: System Sciences: Innovation and Innovation Management, IEEE Computer Society, California [CD Rom, ISBN 0-7695-2507-5], pp. 8.

Fortuin, F.T.J.M. and S.W.F. Omta, 2007a. The Length of the Product Generation Life Cycle as a Moderator of Innovation Strategy: A Comparative Cross-Industry Study of Ten Leading Technology-Based Companies. In: System Sciences: Innovation and Innovation Management, IEEE Computer Society, California, pp. 10, [CD-ROM], ISBN 0-7695-2755-8; ISSN 1530-1605, pp. 10.

Fortuin, F.T.J.M. and S.W.F. Omta, 2007b. Aligning R&D to Business. A Longitudinal Study of BU Customer Value in R&D, International Journal of Innovation and Technology Management, 4 (4), 1-21.

Foss, N.J., 1994. Realism and Evolutionary Economics. Journal of Social and Biological Systems, 17, 21-40.

Foss, N.J., 2005. Scientific Progress in strategic Management: The case of the Resource-Based View. Journal of Learning and Intellectual Capital, Special Issue: 20 Years after the Resource-Based View.

Fredrikson, J.W., 1984. The Comprehensiveness of Decision-making Processes: Extension, Observations, Future Directions. Academy of Management Journal, 27 (3), 445-466.

Freeman, C., 1982. The Economics of Industrial Innovation. Pinter, London.

Freeman, C. and L. Soete, 1997. The Economics of Industrial Innovation. Pinter, London.

Galbraith, J.R. and D.A. Nathanson, 1979. The role of organizational structure and process in strategy implementation. In D. Schendel & C. W. Hofer (Eds.), Strategic Management: A New View of Business Policy and Planning. Little Brown, Boston, Mass, 249-283.

Gaito, J., 1980. Measurement Scales and Statistics: Resurgence of an Old Misconception. Psychological Bulletin, 87, 564-567.

Gerritsma, F. and S.W.F. Omta, 1998. The Content Methodology. Facilitating Performance Measurement by Assessing the Complexity of R&D projects, Management of Technology, Sustainable Development and Eco-efficiency. Elsevier Science Ltd, Orlando, 101-110.

Ghemawat, P., 1991. Commitment: The Dynamics of Strategy. Free Press, New York.

Gilsing, V.A., 2003. Exploration, Exploitation and Co-evolution in Innovation Networks, Erasmus Research Institute of Management PhD Series Research in Management 32.

Ginsberg, A., 1988. Measuring and Modeling Changes in Strategy: Theoretical Foundations and Empirical Directions. Strategic Management Journal, 9, 559-576.

Ginsburg, A. and N. Venkatraman, 1985. Contingency Perspectives of Organizational Strategy: A Critical Review of the Empirical Research. Academy of Management Review, 10, 421-434.

Glass, J.T., I.M. Ensing and J. Desanctis, 2003. Managing the Ties between Central R&D and Business Units. Research.Technology Management (Jan-Feb), 24-31.

Gopalakrishnan, S., 2000. Unraveling the Links between Innovation and Organizational Performance. Journal of High Technology Management Research, 11 (1), 137-153.

Grant, R.M., 1991. The Resource-Based Theory of Competitive Advantage: Implications for Strategy Formulation. California Management Review, 33 (3), 114 – 133.

Grant, R.M., 1996. Toward a Knowledge-based Theory of the Firm. Strategic Management Journal, 17, 109-122.

Griffin, A., 1997. PDMA Research on new Product Development Practices: Updating Trends and Benchmarking Best Practices. Journal of Product Innovation Management, 14, 429-458.

Griffin, A. and A.L. Page, 1993. An Interim Report on Measuring Product Development Success and Failure. Journal of Product Innovation Management, 10 (3), 291-308.

Gulati, R., N. Nohria and A. Zaheer, 2000. Strategic Networks. Strategic Management Journal, 21 (3), 203-215.

Gupta, A.K., S.P. Raj and D.L. Wilemon, 1985. R&D and Marketing Dialogue in High-Tech Firms. Industrial Marketing Management, 14, 289-300.

Hadjimanolis, A., 2000. A Resource-Based View of Innovativeness in Small Firms. Technology Analysis and Strategic Management, 12 (2), 263-281.

Hagedoorn, J., 1993, Understanding the Rationale of Strategic Technology Partnering: Interorganizational Modes of Cooperation and Sectoral Differences. Strategic Management Journal, 14 (5), 371-386.

Hair, J.F. jr., R.E. Anderson, R.L. Tatham and W.C. Black, 1998. Multivariate Data Analysis, 5th edition. Prentice-Hall Inc., New Jersey.

Hamel, G. and C.K. Prahalad, 1989. Strategic Intent. The Best of the Harvard Business Review. Harvard University Press, 187-200.

Hamel, G. and C.K. Prahalad, 1994. Competing for the Future. Harvard Business School Press, Boston, Mass.

Hart, S., E.J. Hultink, N. Tzokas and H. Commandeur, 1998. How Industrial Companies Stear their New Product Development Processes: Empirical Evidence from Dutch and UK Firms, 5th International Product Development Management Conference, Como, Italy.

Hatten, K.J. and D. Schendel, 1977. Heterogeneity within an Industry: Firm Conduct in the U.S. The Journal of Industrial Economics, 26 (2), 97 – 113.

Hauser, J.R., 2001. Metrics Thermostat. Journal of Product Innovation Management, 18 (3), 134-153.

Heene, A. and R. Sanchez, 1996. Competence-Based Strategic Management. John Wiley & Sons, New York etc.

Helfat, C.E. and M.A. Peteraf, 2003. The Dynamic Resource-Based View: Capability Lifecycles. Strategic Management Journal, 24, 997–1010.

Helfat, C.E. and R.S. Raubitschek, 2000. Product Sequencing: Co-evolution of Knowledge, Capabilities and Products. Strategic Management Journal, 21 (10-11), 961–980.

Henderson, R. and J. Cockburn, 1994. Measuring Competence? Exploring Firm Effects in Pharmaceutical Research. Strategic Management Journal, Vol. 15, pp. 63-84.

Henderson, J.C. and N. Venkatraman, 1993. Strategic alignment: Leveraging Information Technology for Organizations. IBM Systems Journal, 32 (1), 4-16.

Henke, J.W., A.R. Krachenberg and T.F. Lyons, 1993. Cross-Functional Teams: Good Concept, Poor Implementation. Journal of Product Innovation Management, 10, 216-229.

Hise, R.T., L. O'Neal, A. Parasuraman and J.U. McNeal, 1990. Marketing/R&D Interaction in New product Development: Implications for New Product Success Rates. Journal of Product Innovation Management, 7, 142-155.

Hodgson, J.M., 1993, Economics and Evolution. University of Michigan Press, Ann Arbor.

Hofer, C.W. and D. Schendel, 1978. Strategy Formulation: Analytical Concepts. West Publishing, St. Paul.

Holland, J.H., 1975. Adaptation in Natural and Artificial Systems. University of Michigan Press, Ann Arbor.

Hollander J., 2002. Improving Performance in Business Development. Genesis a Tool for Product Development Teams. PhD Thesis University of Groningen, The Netherlands.

Huizenga, E.I., 2000. Innovation management: How Frontrunners Stay Ahead. PhD Thesis, University of Maastricht, The Netherlands.

Hunt, S.D., 2000. A General Theory of Competition: Resources, competences, Productivity, and Economic Growth. Sage, Thousand Oaks.

Hüsig, S. and S. Kohn, 2003. Factors Influencing the Front End of the Innovation Process: A Comprehensive Review of Selected NPD and Explorative FFE Studies, University of Regensburg and Fraunhofer Technologie Entwicklungsgruppe, Regensburg and Stuttgart, Germany.

Huston, L. and N. Sakab, 2006. Connect and Develop: Inside Procter & Gamble's New Model for Innovation. Harvard Business Review, March, 60-72.

Jamrog, J.J., 2006. The Quest for Innovation: A Global Study of Innovation Management 2006-2016, Human Resource Institute, University of Tampa, Tampa, Fl.

Janszen, F.H.A., 2000. The Age of Innovation. Making Business Creativity a Competence, not a Coincidence. Pearson Education London etc.

Janszen, F.H.A., M.P.F. Vloemans and S.W.F. Omta, 1999. Analyzing, Modeling and Simulating the New Business and Market Development Process, Portland International Conference of the Management of Engineering and Technology, Portland, Oregon.

Jaruzelski, B., K. Dehoff and R. Bordia, 2005. Money Isn't Everything. Strategy + Business, Winter, 54-67.

Johne, A. and P. Snelson, 1988. Success Factors in Product Innovation: A Selected Review of the Literature. Journal of Product Innovation Management, .5 (2), 100-110.

Johnson, G. and Scholes, K., 2002. Exploring Corporate Strategy. Prentice Hall, New York.

Judge, W.Q. and A. Miller, 1991. Antecedents and Outcomes of Decision speed in Different Environments. Academy of Management Journal, 34 (2), 449–464.

Kamath, R.R. and J.K. Liker, 1990. Supplier Dependence and Innovation: A Contingency Model of Supplier's Innovative Activities. Journal of Engineering and Technology Management, 7, 111-127.

Katsikeas, C.S., S. Samiee and M. Theodosiou, 2006. Strategy Fit and Performance Consequences of International Marketing Standardization. Strategic Management Journal, 27 (9), 867-890.

Katz, D. and R.L. Kahn, 1978. The Social Psychology of Organizations. John Wiley & Sons, New York, London.

Kim, L., 1999. Building Technological Capability for Industrialization: Analytical Frameworks and Korea's Experience. Industrial and Corporate Change, 8 (1), 111–136.

King, W.R., 1978. Strategic Planning for Management Information Systems. MIS Quarterly, 2 (1), 27-37.

Kline, S.J. and N. Rosenberg, 1986. An Overview of Innovation. In: R. and N. Rosenberg-Landau (Eds.), National Academic Press, Washington DC, 275-303.

Kogut, B. and U. Zander, 1992. Knowledge of The Firm, Combinative Capabilities, and the Replication of Technology. Organization Science, 3(3), 383-397.

Kuczmarski and Associates, Inc., 1994, Winning New Product and Service Practices for the 1990's. Kuczmarski and Associates, Chicago.

Langlois, R.N., 1986, Economics as a Process: Essays in the New Institutional Economics. Cambridge University Press, Cambridge UK.

Lawrence, P.R. and J.W. Lorsch, 1967. Organization and Environment: Managing Differentiation and Integration. Harvard University Press, Boston, Mass.

Lefebvre, L.A., R. Mason and E. Lefebvre, 1997. The Influence Prism in SME's. The Power of CEOs' Perceptions on Technology Policy and Its Organizational Impacts. Management Science 43 (6), 856-878.

Lemak, D.J. and W. Arunthanes, 1997. Global Business Strategy: A Contingency Approach. Multinational Business Review 1, 26–38.

Leonard-Barton, D., 1995, Wellsprings of Knowledge. Building and Sustaining the Sources of Innovation. Harvard Business School Press, Boston, Mass.

Levinthal, D.A. and J.G. March, 1993. The Myopia of Learning. Strategic Management Journal, 14 (Winter Special Issue), 95-112.

Levitt, T., 1965, Exploit the Product Life Cycle. Harvard Business Review (November-December) 43, 81-96.

Liker, J.K. and W.M. Hancock, 1986, Organization Systems Barriers to Engineering Effectiveness, IEEE Transaction on Engineering Management EM-33 (2), 82-91.

Lindblom, C.E., 1959. The Science of 'Muddling Through'. Public Administration Quarterly, 19, 79-88.

Lippman, S.A. and R.P. Rumelt, 1982. Uncertain Imitability: An Analysis of Interfirm Differences in Efficiency under Competition. Bell Journal of Economics, 13 (2), 418-438.

Lippman, S.A. and R.P. Rumelt, 1992. Demand Uncertainty and Investment in Industry-Specific Capital. Industrial and Corporate Change, 1 (1), 235-262.

Luehrman, T.A., 1998. Strategy as a Portfolio of real Options. Harvard Business Review, September-October, 89-99.

Lukas, B.A., J.J. Tan and J.T.M. Hult, 2001. Strategic Fit in Transitional Economies: The Case of China's Electronics Industry. Journal of Management 27, 409–429.

Luo, Y. and Peng, M.W., 1999. Learning to Compete in a Transition Economy: Experience, Environment and Performance. Journal of International Business Studies 34 (3), 290-309.

Maidique, M.A. and B.J. Zirger, 1985. The New Product Learning Cycle. Research Policy 14, 299-313.

March, J.G., 1991. Exploration and Exploitation in Organizational Learning. Organization Science, 2 (1), 71-87.

Martin, M.J.C., 1985. Managing Technological Innovation and Entrepreneurship. Reston Publishing Company, Reston, Virginia.

Mason, E.S., 1939. Price and Production Policies of Large Scale Enterprises. American Economic Review, 29 (1), 61 – 74.

McCarthy, M., 2003. Linking Technological Change to Business Needs. Research.Technology. Management, 46 (2), 47-52.

McGrath, M.E., 1995. Product Strategy for High-Technology Companies: How to Achieve Growth, Competitive Advantage, and Increased Profits. McGraw-Hill, New York.

McGrath, M.E. (Ed.) 1996. Setting the Pace in Product Development. A Guide to Product and Cycle-Time Excellence. Butterworth-Heinemann, Boston etc.

McGrath, M.E. and M.N. Romeri, 1994. From Experience: The R&D Effectiveness Index: A Metric for Product Development Performance. Journal of Product Innovation Management, 11 (3), 213-220.

Mercer Management Consulting, Inc., 1994. High Performance New Product Development: Practices That Set Leaders Apart. Mercer Management Consulting, Inc., Boston, Mass.

Meyer, M.H. and J.M. Utterback, 1993. The Product Family and the Dynamics of Innovation. Sloan Management Review, 34 (3), 29-47.

Miles, R.D. and C.C. Snow, 1978, Organizational Strategy, Structure, and Process. McGraw-Hill, New York.

Miles, R.E. and C.C. Snow, 1994. Fit, Failure and the Hall of Fame. Free Press, New York.

Milgrom, P. and J. Robberts, 1990. The Economics of Modern Manufacturing: Technology, Strategy, and Organization. American Economic Review, 80 (3), 511-528.

Miller, D., 1992. Environmental Fit versus Internal Fit. Organization Science, 3 (2), 159-178.

Miller, D. and J. Shamsie, 1996. The Resource-Based View of the Firm in Two Environments: The Hollywood Film Studies from 1936 to 1965. Academy of Management Journal, 39 (3), 519-543.

Miller, W.L. and L. Morris, 1999. Fourth Generation R&D: Managing Knowledge, Technology, and Innovation. John Wiley & Sons, New York etc.

Mintzberg, H., 1979. The Structuring of Organizations: A Synthesis of the Research. Prentice-Hall, Engelwood Cliffs, New Jersey.

Mintzberg, H., 1994. The Rise and Fall of Strategic Planning. Prentice Hall, Hemel Hempstead.

Mintzberg, H. and J. Lampbel, 1999. Reflecting on the strategy Process. MIT Sloan Management Review, 40, 21-30.

Mintzberg, H. and J.B. Quinn, 1991. The Strategy Process: Concepts, Contexts, Cases. Prentice Hall, Englewood Cliffs, New Jersey.

Moncrieff, J., 1999. Is Strategy Making a Difference? Long Range Planning, 32 (2), 273-276.

Moore, G.A., 1995. Crossing the Chasm. Marketing and Selling High-techProducts to Mainstream Customers. HarperBusiness, New York.

Moss Kanter, R.M., J. Kao and F. Wiersema, 1997. Innovation: Breakthrough Thinking at 3M, DuPont, GE, Pfizer, Rubbermaid. Harper Collins Publishers, New York.

Nelson, R.R., 1991. Why do Firms Differ, and How does it Matter? Strategic Management Journal, Winter Special Issue (12), 61-74.

Nelson, R.R. (Ed.), 1993. National Systems of Innovation. Oxford University Press, Oxford.

Nelson, R.R. and S.G. Winter, 1982. An Evolutionary Theory of Economic Change. Harvard University Press, Cambridge, Mass.

Niosi, J., 1995. Flexible Innovation: Technological alliances in Canadian Industry. Mc Gill-Queens University Press, Montreal.

Nonaka, I., 1994. A Dynamic Theory of Organizational Knowledge Creation. Organization Science, 5 (1), 14-37.

Norton, J.M., E. Parry and X.M. Song, 1994. Integrating R&D and Marketing: A Comparison of Practises in the Japanese and American Chemical Industries. IEEE Transactions on Engineering Management, 41 (1), 5-20.

Nunnally, J.C., and I.J. Bernstein, 1994. Psychometric Theory, Third Edition. McGraw-Hill, New York.

OECD, 1994. The Measurement of Scientific Activities: Proposed Standard Practice for Surveys of Research and Experimental Development. The Frascati Manual, Organization for Economic Cooperation and Development, Paris.

Omta, S.W.F., 1995. Critical Success Factors in Biomedical Research and Pharmaceutical Innovation. Kluwer Academic Publishers, Dordrecht, London, Boston.

Omta, S.W.F. and J.M.L. Van Engelen, 1998. Preparing for the 21st Century, Research. Technology. Management, 41 (1), 31-44.

Omta, S.W.F. and J. Bras, 2000. Design and Implementation of the Balanced Score Card in Industrial R&D. In: T. M. Khalil and L. Lefebvre (Eds.), 9th International Conference on the Management of Technology, Miami.

Omta, S.W.F. and P. Folstar, 2005. Integration of Innovation in the Corporate Strategy of Agri-food Companies, Innovation in Agri-Food Systems. Wageningen Academic Publishers, Wageningen, 223-246.

Oppenheim, A.N., 1966. Questionnaire Design and Attitude Measurement, Basic Books, New York.

Park, S. and Gil, Y., 2006. How Samsung Transformed its Corporate R&D center. Research.Technology Management, 49 (4): 24-29.

Patterson, M.L., 1993. Accelerating Innovation: Improving the Process of Product Development. Van Nostrand Reinhold, New York.

Peng, M.W. and A.S. York, 2001. Behind Intermediary Performance in Export Trade: Transactions, Agents and Resources. Journal of International Business Studies 31 (4), 535-554.

Penrose, E.T., 1959. The Theory of the Growth of the Firm. Basill Blackwell, Oxford.

Peteraf, M., 1993. The Cornerstones of Competitive Advantage: A Resource-Based View. Strategic Management Journal, 14 (3), 179-191.

Pettigrew, A.M., 1990. Longitudinal Field Research on Change: Theory and Practice. Organization Science, 1 (3), 267-292.

Pittiglio, Rabin, Todd and McGrath, 1995. Product Development Leadership for Technology-Based Companies: Measurement and Management -A Prelude to Action. Pittiglio, Rabin, Todd and McGrath (PRTM), Weston, Mass.

Polli, R. and V. Cook, 1969. Validity of the Product Life Cycle. Journal of Business (Oct.) 42, pp. 385-400.

Porter, M.E., 1980. Competitive Strategy: Techniques for Analyzing Industries and Competitors. Free Press, New York.

Porter, M.E., 1985. Technology and Competitive Advantage. In: M. E. Porter (Ed.), Competitive Advantege: Creating and Sustaining Superior Performance. The Free Press, New York, 164-200.

Porter, M.E., 1998. On Competition. Free Press, New York.

Prahalad, C.K. and G. Hamel, 1990. The Core Competence of the Corporation. Harvard Business Review, 68, 79-91.

Quinn, J.B., 1980. Strategies Through Change: Logical Incrementalism. Irwin, Homewood, Ill.

Quinn, J.B., 2000. Outsourcing Innovation: The New Engine of Growth. Sloan Management Review, Summer, 13-28.

Raz, T., A. Shenhar and D. Dvir, 2002. Risk Management, Project Success, and Technological Uncertainty, R&D Management, 32 (2), 101-109.

Reid, S.E. and U. De Brentani, 2004. The Fuzzy Front End of New Product Development for Discontinuous Innovations: A Theoretical Model. Journal of Product Innovation Management, 21, 170-184.

Roberts, E.B., 1995. Benchmarking the Strategic Management of Technology, Research.Technology. Management (Jan-Feb), 44-56.

Rochford, L. and W. Rudelius, 1992. How Involving More Functional Areas within a Firm Affects the New Product Process. Journal of Product Innovation Management, 9, 287-299.

Rogers, E.M., 1995. The Diffusion of Innovation, 4th ed. The Free Press, New York.

Rosenau, M.D. jr, A. Griffin, G.A. Castellion and N.F. Anschuetz (Eds.), 1996. The PDMA Handbook of New Product Development. John Wiley & Sons, New York etc.

Rosenberg, N., 1982. Inside the Black Box: Technology and Economics. Cambridge University Press, Cambridge.

Rosenbloom, R. and R.A. Burgelman, 1989. Technology Strategy: An Evolutionary Process Perspective. Research on Technological Innovation, Management and Policy, 4.

Rosner, M.M., 1968. Economic Determinants of Organizational Innovation. Administrative Science Quarterly, 12 (4), 614-625.

Rothwell, R., 1993. The Fifth Generation Innovation Process. In: Oppenländer, K.-H. (Ed.). Privates und Staatliches Innovationsmanagement, München, Germany.

Roussel, P.A., K.A. Saad and T.J. Erickson, 1991, Third Generation R&D. Harvard Business School Press, Boston, Mass.

Rumelt, R.P., 1982. Diversification Strategy and Profitability. Strategic Management Journal, 3, 359-369.

Rumelt, R.P., 1987. Theory, Strategy, and Entrepreneurship. In: D.J. Teece (Ed.), The Competitive Challenge - Strategies for Industrial Innovation and Renewal. Ballinger, Cambridge Mass., 137-158.

Sanchez, R., 2001. Building Blocks for Strategy Theory: Resources, Dynamic Capabilities and Competences. In: H.W. Volberda and T. Elfring (Eds.), Rethinking Strategy. Sage Publications, London etc., 143-158.

Sanchez, R., A. Heene and H. Thomas, 1996. Dynamics of Competence-Based Competition. Elsevier Science, New York.

Sanchez, R. and A. Heene, 1997. Strategic Learning and Knowledge Management. John Wiley & Sons, New York etc.

Sanderson, S.W. and M. Uzumeri, 1997. The Innovation Imperative: Strategies Managing Product Models and Families. Irwin Professional Publishing, Chicago.

Schendel, D.E. and G.R. Patton, 1978. A Simultaneous Equasion Model of Corporate Strategy. Management Science, 24, 1611 1621.

Scholten, V.F., 2006. The Early Growth of Academic Spin-offs. PhD Thesis Wageningen University, The Netherlands.

Schumpeter, J.A., 1934. Theory of Economic Development. Harvard University Press, Cambridge, Mass.

Schumpeter, J.A., 1942. Capitalism, Socialism and Democracy. Harper and Row, New York.

Schwartz, H. and S.M. Davis, 1981. Matching Corporate Culture and Business Strategy. Organizational Dynamics, 10 (1), 30-48.

Scot, M., 1991. Longitudinal Research. Sage, Newbury Park, CA.

Selznick, P., 1957. Leadership and Administration. Harper and Row, New York.

Senge, P.M., 1990. The Fifth Discipline. The Art and Practice of the Learning Organization. Currency Doubleday, New York.

Shane, S.A. and K.T. Ulrich, 2004. Technological Innovation, Product development, and Entrepreneurship in Management Science. Management Science 50 (2), 133-144.

Shenhar, A.J. and A. Tishler, 2002. Refining the Search for Project Success Factors: A Multivariate, Typological Approach. R&D Management, 32 (2), 111-124.

Shrivastava, P., A. Huff and J.E. Dutton (Eds.), 1992. Advances in Strategic Management. Jai Press, Greenwich.

Shuen, A., 1994. Technology Sourcing and Learning Strategies in the Semiconductor Industry, unpublished Ph.D. dissertation, University of California, Berkeley.

Smith, P.G. and D.G. Reinertsen, 1998. Developing Products in Half the Time. John Wiley & Sons, New York etc.

Smith, D., 2006. Exploring Innovation. McGraw-Hill, London etc.

Song, X.M. and M.E. Parry, 1992. The R&D-Marketing Interface in Japanese High-Technology Firms. Journal of Product Innovation Management, 9, 91-112.

Souder, W.E., 1981. Disharmony between R&D and Marketing. Industrial Marketing Management, 10, 67-73.

Stevens, S.S., 1946. On the Theory of Scales of Measurement. Science, 103, 677-680.

Stevens, S.S., 1951. Mathematics, Measurement, and Psychophysics. In S.S. Stevens, (Ed.), Handbook of Experimental Psychology. John Wiley & Sons, New York etc.

Stevens, G.A. and J. Burley, 1997. 3,000 Raw Ideas = 1 Commercial Success! Research. Technology Management 40 (3), 16-27.

Storey, J. (Ed.), 2004. The Management of Innovation. Edward Elgar Publishing, Cheltenham.

Subramanian, A., 1996. Innovativeness: Redefining the Concept. Journal of Engineering and Technology Management, 13, 223-243.

Teece, D.J. and G. Pisano, 1994. The Dynamic Capabilities of Firms: An Introduction. Industrial and Corporate Change, 3 (3), 537-556.

Teece, D.J., G. Pisano and A. Scheun, 1997. Dynamic Capabilities and Strategic Management. Strategic Management Journal, 18 (7), 509-533.

Thompson, J.D., 1967. Organizations in Action. McGraw-Hill, New York.

Tidd, J., J. Bessant and K. Pavitt, 2001. Managing Innovation: Integrating Technological, Market, and Organizational Change. John Wiley & Sons, New York etc.

Troy, K.L., 2004. Making Innovation Work: From Strategy to Practice, The Conference Board.

Truijens, O., 2004. Towards a Theory of Information Strategy: Exploiting Market Opaqueness in Search for InfoRent, University of Amsterdam.

Twiss, B.C., 1992. Managing Technological Innovation. Pitman Publishers, London.

UK R&D Scoreboard 1996, 1997, 2005, 2006. Department of Trade and Industry, Company Reporting Ltd. The International Ranking of Top 300/1000 companies by R&D Expenditure (www.innovation.gov.uk).

Van de Ven, A.H., H.L. Angle and M.S. Poole (Eds.), 1989. Research on the Management of Innovation. Ballinger/Harper & Row, New York.

Van de Ven, A.H., 1990. Methods for studying Innovation Development in the Minnesota Research Program. Organization Science, 1 (3), 313-335.

Venkatraman, N. and J.C. Camillus, 1984. Exploring the Concept of Fit and Technology Management, 30 (1-2), 131-146.

Venkatraman, N., 1989. The Concept of Fit in Strategy Research. Toward Verbal and Statistical Correspondence. Academy of Management Review 14 (3), 423–444.

Venkatraman, N. and J.E. Prescott, 1990. Environment-Strategy Coalignment.: An Empirical Test of its Performance Implications. Strategic Management Journal 11 (1), 1–23.

Verdú Jover, A.J., J.F. Lloréns Montes and V.J. García Morales, 2005. Flexibility, Fit and Innovative Capacity: An Empirical Examination. International Journal of Technology Management, 30 (1-2), 131-146.

Volberda, H.W., 1992. Organizational Flexibility. Change and Preservation. PhD Thesis University of Groningen, The Netherlands.

Von Hippel, E., 1988. The Sources of Innovation. Oxford University Press, Oxford.

Von Zedtwitz, M., 1999. Managing Interfaces in International R&D. PhD Thesis University of St. Gallen nr 2315. Switzerland.

Wernerfelt, B., 1984. A Resource-Based View of the Firm. Strategic Management Journal, 5, 171-180.

Williams, J.R., 1998. Renewable Advantage. Crafting Strategy through Economic Time. The Free Press, New York.

Wheelwright, S.C. and K.B. Clark, 1992. Revolutionizing Product Development: Quantum Leaps in Speed, Efficiency, and Quality. The Free Press, New York.

Winter, S.G., 2003. Understanding Dynamic Capabilities. Strategic Management Journal, 24, 991–995.

Witt, U., 1992. Evolutionary Concepts in Economics. Eastern Journal of Marketing, 18 (4), 405-419.

Yamada, A. and C. Watanabe, 2005. Firms with adaptability lead the way to innovative development. Proceedings of the 14th International Conference on Management of Technology, ed.: T. Khalil, Vienna, May 22-26.

Yin, R.K., 1989. Case Study Research-Design and Methods. Sage Publications, London.

Zaltman, G., R. Duncan and J. Holbeck, 1973. Innovation and Organization. John Wiley & Sons, New York etc.

Zajac, E.J., M.S. Kraatz, and R.K.F. Bresser, 2000. Modeling the Dynamics of Strategic Fit: A Normative Approach to Strategic Change. Strategic Management Journal, 21, 429-453.

Zbignew, J. and F. Pasek, 2002. Linking Strategic Planning with R&D Portfolio Management in an Engineering Research Centre, 5th International Conference on Managing Innovation in Manufacturing. Milwaukee, Wisconsin.

Zollo, M. and S.G. Winter, 2002. Deliberate Learning and the Evolution of Dynamic Capabilities. Organization Science, 13 (3), 339–351.

Appendices

Appendix A. General questions in the cross-industry study

1 Strategic alignment

- What assessment tools does the company use to identify promising areas for innovation (scenario-based planning, technology and market road maps etc.)?

2 Degree of exploration

- Does your laboratory primarily concentrate on basic technology development or on applied development tasks?
- What are the most important improvement goals for the coming years, e.g. customer focus, cycle time reduction etc.
- What is the company's policy on patenting? What kind of patent screening process do you use, and what kind of patent assessment systems are in place?

3 R&D competencies

3.1 In-house competencies

- What are the key innovation areas for your laboratory for the coming years?
- What are the key technologies and the core technology competencies of your company?
- Which competencies should be strengthened to compete in the future?

3.2 Open innovation

- What is your lab's policy on R&D cooperation (i.e. sponsoring and contracting out to universities and/or to institutes, cooperation with suppliers and buyers, strategic alliances and joint ventures)?

4 R&D capabilities

- What are the core management systems in your laboratory?
- What competitor intelligence assessment systems are in place?
- Do you use expert systems to assess the tacit knowledge of your staff?

4.1 Flexibility

- Would it be possible to shorten the time-span of the different steps in the R&D process?

4.2 Internal communication

- What kinds of information and communication technology (ICT) do you use (video conferences, electronic meeting rooms etc.? What is your experience with these?
- Do you make R&D data available in-house, for instance via shared data bases, electronic discussion forums or via the intranet?

- What is your policy toward ICT security, especially for multi-site platforms? Do you use limited access, authentication and/or data encryption?
- Which functions are present in the project teams of major projects, e.g. marketing, manufacturing, purchasing, law, and/or environmental experts?

4.3 External communication
- Do you make non-critical R&D data available to the general public, e.g. via the Internet?
- What is the company's policy towards scientific publishing?

4.4 Incentive systems
- How can the primary and secondary working conditions in your laboratory be assessed in comparison with other technology-based companies?
- What incentives are given to scientific staff (results-based compensation, both material and immaterial)?
- Is there a dual (or hybrid) career system (managerial and scientific)?
- How is innovation stimulated in your company (i.e. awards, funds, recognition and/or fellowships)?

5 R&D throughput
- Who is responsible for stage gate review and what criteria determine the initiating, changing or phasing out of development projects?
- How is your technology transferred to the divisions?

6 Methods to improve strategic alignment

6.1 R&D funding
- To what extent are you funded by the business units? What are the positive and negative effects of BU funding?
- To what extent are you externally funded and are these funds effective?

6.2 Management methods
- How is the emphasis on customer value embedded in R&D (customer focus groups etc.)?
- Do you use tools to improve the links to business, such as Quality Function Deployment, value engineering, and DFX – design for Manufacturing, Assembly and Service.
- Which technical and management methods are used to shorten the time-to-market, e.g. rapid prototyping, virtual development (modeling and simulation) and concurrent engineering?

Appendix B. Cross-industry survey questionnaire I

1. Please indicate roughly the 1995, 1996 and 1997 figures for your company

	1995	1996	1997
Worldwide sales (billion $)	....	....	....
Number of employees	....	....	....
Number of divisions	....	....	....
Operating profit margin (%)	....	....	....
Average number of employees per division	....	....	....
Total R&D spending ($m)	....	....	....
Number of R&D employees	....	....	....

2. Please name the three most important product divisions with their percentage of sales revenue

Product group 1 .... Sales revenue .... %
Product group 2 .... Sales revenue .... %
Product group 3 .... Sales revenue .... %

3. Please indicate roughly the average product generation life cycle of a typical product in:

Product group 1 .... years Product group 3 .... years
Product group 2 .... years

4. Please indicate roughly the total revenue contribution of new products introduced in the last three years for the company as a whole, and for the three most important product groups:

Company .... % Product group 2 .... %
Product group 1 .... % Product group 3 .... %

5a. Please estimate roughly the number of patents that your company obtained through the efforts of your R&D department over the past three years .... patents

5b. Please estimate roughly what percentage of patents result in new products and processes introduced on to the market: .... %

5c. Please estimate roughly what percentage of your annual R&D budget is recovered by royalties on the basis of out licensing: .... %

5d. Please estimate roughly what percentage of patentable ideas actually lead to patents being granted: .... %

5e. Please estimate roughly what percentage of sales is protected by patents owned by the company: .... %

5f. Does your company (R&D department and/or divisions) monitor the patent portfolios of your main competitors on a regular basis? Yes/No

6. Please estimate roughly what percentage of savings in the costs of products and processes sold in 1997 comes from product and process changes adopted in the preceding three years: .... %

7. What percentage of 2000 sales do you expect to be generated from projects in your R&D pipeline: .... %

8. Please indicate roughly the number of staff per function in your R&D department in fulltime equivalents (ftes):

	Scientists/engineers	Technical Support staff
Directorate/management team	.... ftes	.... ftes
Research and technology development	.... ftes	.... ftes
Product and process development	.... ftes	.... ftes
Engineering and testing	.... ftes	.... ftes
Technical & administration services	.... ftes	.... ftes

9. Please indicate roughly the percentage of total R&D spending which is directed towards:

Research and technology development	.........%
Product and process development	.........%
Engineering and testing	.........%
Design tools	.........%

10. Please indicate roughly the number of scientists and engineers and technical support staff and the percentage of R&D spending which is directed towards:

	Scientists/Engineers	Support Staff	R&D spending
Pioneering projects	.... ftes	.... ftes	.... %
Major projects	.... ftes	.... ftes	.... %
Minor projects	.... ftes	.... ftes	.... %

11. Please indicate roughly the educational level of the scientists and engineers in your R&D department:

    PhD .... ftes MSc .... ftes Higher Vocational Education .... ftes

12. Please indicate roughly the percentage of business division funding in your R&D budget: .... %

13. Please indicate roughly the percentage of total R&D which is conducted:

In-house	.... %
By contractors/suppliers	.... %
In universities and institutes (contract research and sponsoring)	.... %
In R&D cooperation (strategic alliance, joint venture etc.)	.... %

14. Please mention your most important partners in outsourcing, sponsoring and R&D cooperation:

15. Please mention your strategy on outsourcing, sponsoring and cooperation:

16. Please indicate roughly the time spent and the percentage of R&D spending per R&D phase for a typical major project:

	Time spent	R&D spending
Concept development	...... months	...... %
Specification/planning	...... months	...... %
Product/process engineering	...... months	...... %
Release to manufacturing	...... months	...... %
Total	...... months	100 %

17. Please indicate roughly for each phase of the R&D process the percentage of R&D projects that are stopped in that phase and the percentage of R&D projects which that are still being conducted according to schedule:

	Projects stopped	Projects on schedule
Concept development	...... %	...... %
Specification/planning	...... %	...... %
Product/process engineering	...... %	...... %
Release to manufacturing	...... %	...... %

18. If we normalize the current average project cycle time at 100, what was the average project cycle time in 1994, and what is the goal for the year 2000?

1994 = ...	1997 = 100	2000 = ...

19. Please mention the main measures you take to shorten the project cycle time:

20. Please rank the goals listed below in accordance with their relative importance to the company (1 = most important goal, 2 = second most important etc.) and estimate the percentage of major projects that meet their original goals (set in the specification and planning phase):

	Rank	R&D projects meeting the goals
Contribution to profitability	....	.... %
Cost reduction	....	.... %
Timeliness of delivery	....	.... %
Within planned budget	....	.... %
Performance to specifications	....	.... %
Performance to customer needs	....	.... %
Other goals, please specify	....	.... %

21. Please characterize the project teams in your company:

	Pioneering projects	Major projects
No. of project teams	....	....
Project team size	....	....
Full-time project leader (%)	....	....
No. of projects per team member	....	....

22. Please indicate roughly the percentage of pioneering and major projects in which staff members of the following departments/companies are involved:

Marketing	..... %	Manufacturing	..... %
Finance	..... %	Major customers	..... %
Major suppliers	..... %	Other companies	..... %
Others, please specify	..... %		

23a. Please estimate how many international conferences, symposia or seminars were attended per R&D staff member: ...... conferences/staff member

23b. How many papers did they present there? ...... papers/staff member

23c. Please estimate how many scientific/technical papers were published in international scientific/technical journals: ...... papers/staff member

23d. Please indicate roughly how many colleagues from other companies abroad visited your R&D department during 1997: ...... colleagues

24. Please indicate roughly the percentage of international meetings within your R&D department (for instance, with the business divisions) that are conducted via video conferencing: ...... %

25a. How many networks does your R&D department have? ....

25b. Is the building of networks stimulated by your company? Yes/No

25c. If so, how is the building of networks stimulated?

Monitoring issues where new networks could be formed	Yes/No
Company-wide publishing of experiences of existing networks	Yes/No
Providing assistance by forming new networks	Yes/No
Other, please specify	

25d. What are the main objectives of these networks?

26. Please indicate roughly the frequency of project team meetings:

1. (Less than) once a week. 2. Every 1 to 2 weeks. 3. 2 weeks to 1 month. 4. 1 to 3 months. 5. (Less than) 3 months

27. How would you characterize your company's project review system?

A calendar-based status review ......
A phase review system at key decision points (Go/No-Go moments) ......
A combination of both ......

28. How often do you review your R&D projects?

1. Once a month. 2. Every quarter. 3. Biannually. 4. Annually. 5. No formal review

29. Please rank (1 = first responsible, 2 = second etc.) those who are responsible for starting, stopping or changing R&D projects:

	Start	Change	Stop
Project (program) leader	....	....	....
Project team members	....	....	....
Functional (department) manager	....	....	....
R&D management team	....	....	....
Cross-functional management team	....	....	....
Cross-business team	....	....	....
Other, please specify	....	....	....

30. The R&D staff reviews the R&D project portfolio with the:

Divisions		Major customers	
Every 3 months	.....	Every 3 months	.....
Every 6 months	.....	Every 6 months	.....
Once a year	.....	Once a year	.....
No formal meetings	.....	No formal meetings	.....

31. The main objectives of these meetings are: (Please rank 1 = most important objective)

To assess the quality of R&D products and processes .....
To ensure that the project portfolio is in line with what they want .....
To find out what products, processes or technologies are needed in the future .....
Other objectives, please specify .....

32. Please indicate roughly the number of hours per R&D staff member spent in direct contact with external customers in 1997 .....

33. What is the frequency of R&D process audits (peer reviews)?

1. Biannually. 2. Annually. 3. Once every 3 years. 4. Once every 5 years. 5. No formal audits

34. Who is involved in R&D process audits?

Senior management ..... Major customers ...... Technical experts .....
Others, please specify .....

35. If an audit showed that a large part (e.g. 20%) of the personnel and material means should be allocated to a new R&D area, how long would it take for this reallocation to be realized?

    1. Less than 1 month. 2. 1 to 3 months. 3. 3 to 6 months. 4. 6 to 12 months. 5. More than a year

36. Please indicate roughly the typical time-span between the request and the approval for:

Appointment of an engineer		Purchase of apparatus > US$ 50,000	
Less than 1 week	.....	Less than 1 week	.....
1 week to 1 month	.....	1 week to 1 months	.....
1 to 3 months	.....	1 to 3 months	.....
3 to 6 months	.....	3 to 6 months	.....
More than 6 months	.....	More than 6 months	.....

37. From the date of a request from business divisions, a typical R&D project will start within: (Please tick one of the boxes for major projects, one of the boxes for minor projects, and one of the boxes for extended tasks)

Major projects (> 50 man-weeks)		Minor projects (7 to 50 man-weeks)		Extended tasks (2 to 7 man- weeks)	
3 months	.....	1 month	.....	2 weeks	.....
6 months	.....	3 months	.....	1 month	.....
1 year	.....	6 months	.....	3 months	.....
> 1 year	.....	> 6 months	.....	> 6 months	.....

38. Please indicate roughly the staff flow in 1997 (in ftes):

    From R&D to other departments .... ftes
    From R&D to other organizations (companies, universities, institutes) .... ftes

39. On average, what percentage of the scientists and engineers were not in-house due to:

Training programs and apprenticeships	.... %
International meetings	.... %
Vacation	.... %
Illness	.... %
Other, please specify	.... %

40. Please indicate which of the incentives mentioned below are in place in your company.

Material incentives	
Salary level in comparison to competitors	Higher/equal/lower
Stock options for R&D staff	Yes/No
Company car for R&D staff	Yes/No
Patent profit sharing	Yes/No
Extra payment for extraordinary research efforts	Yes/No
Immaterial incentives	
R&D staff may spend part of their work on own projects	Yes/No
Dual (or hybrid) ladder system (managerial and scientific)	Yes/No
Stimulation of scientific publishing and presentation	Yes/No
Innovation awards	Yes/No
In-company innovation capital funds	Yes/No
Technology society	Yes/No
Fellowships	Yes/No
Others, please specify	

Appendix C. Cross-industry survey questionnaire II

Please circle for the questions below the figure (1 to 7 inclusive) which best reflects your opinion.

Not true at all 1 2 3 4 5 6 7 *Totally true*

1. Market and technology trends are important inputs to product strategy; scenarios (including trend-breaking scenarios) are developed to assess technology roadmaps with short-, medium- and long-term time horizons.

2. Technology roadmaps are updated regularly to reflect relevant changes in circumstances (e.g. technology breakthroughs or unexpected problems).

3. The R&D project portfolio is based upon a strategic vision for intended future products and relevant target markets.

4. The core technologies are well defined.

5. There is a regular cross-functional screening process to identity, document, evaluate, and select new product/technology opportunities.

6. The international R&D capability of the company is exploited efficiently, e.g. through the exchange and sharing of experiences and best practices and through the use of multinational product development teams.

7. Possible relationships with third parties and with customers (e.g. for co-design and co-development) are explicitly taken into account in the execution of R&D projects.

8. Possible gaps between current availability and potential future project requirements (skills, resources, and infrastructure) are explicitly established at the start of the project and updated regularly.

9. The cost drivers and capital constraints of the products and processes (e.g. technical and business risks and manufacturability) as well as of the R&D process (e.g. investment in new development tools and manufacturing capability) are well understood and are taken into account at the start of the project and during development.

10. We continually look for cost-saving opportunities while maintaining or improving customer value.

11. Structured design methodologies such as Design For Manufacturing, and Design for Service ensure constant attention to product or process robustness and cost-effectiveness.

12. Effective communication structures (e.g. project progress reports and regular information meetings) facilitate communication with the divisions.

13. Current market information (such as segmentation, trends and feedback on competitors' products and processes) is passed on by marketing to R&D on a regular basis.

14. Current information on competitors' activities (e.g. market performance and customer perception) is passed on by marketing to R&D on a regular basis.

15. Information meetings for R&D staff and customer-oriented staff (such as staff from marketing, sales and service) are held on a regular basis.

16. New product forums involving R&D staff and customers are held regularly.

17. Ideas for new products and processes are evaluated in terms of their value to customers. Customer benefits and drawbacks are the primary input to R&D project prioritizing and to the formulation of specifications.

18. Structured tools such as Quality Function Deployment are used to translate customer requirements into product specifications.

19. R&D staff actively participates in the establishment of the divisions' business plans.

20. R&D project objectives are set in line with divisions' business plans (e.g. concerning new product and market opportunities).

21. R&D project evaluation explicitly refers to alignment with divisions' business plans.

22. The key R&D project parameters (e.g. objectives, limiting factors and verifiable milestones) are discussed with the divisions and, if necessary, reformulated.

23. The progress of R&D projects against milestones is monitored regularly. Any deviation from the original plan is immediately addressed.

24. The progress of all R&D projects is communicated regularly to the relevant divisions.

25. Corporate executives and senior managers are troubled by the question: What are we getting for our R&D investment?

26. Senior management devotes little attention to the early stages of new product development.

27. BU managers are involved in R&D planning.

28. R&D strategy is closely linked to business strategy.

29. Speed is emphasized over budgets in product development.

30. Great emphasis id placed on developing linkages with other organizations for successful new product and process development.

31. There is excellent communication between R&D and marketing.

32. There is excellent communication between R&D and manufacturing.

33. R&D interacts extensively with customers.

34. The use of contract research through independent R&D firms is increasing.

35. Finding funding for high-risk basic research is becoming more difficult.

36. There is a growing realization that incremental technical advances rather than breakthroughs are critical to turning R&D into profitable products and processes.

37. Our suppliers are involved early in the product development process.

38. Senior management forces the divisions to generate a considerable amount (i.e. 25%) of annual sales from new products and services.

39. Each division selects 1 to 3 high-priority R&D project(s) which are developed in a joint effort with R&D.

40. Few restrictions are imposed on the R&D department by administrative regulations (e.g. regarding travel, budget, etc.).

Appendix D. Longitudinal survey questionnaire

In the next two questions (1a and 1b) please choose one of the listed alternatives and enter its number (1 to 5).

1a. Please indicate the relative importance of each of the following objectives for the corporate R&D center.

Alternatives: 1. = an unimportant objective; 2. = an objective of less importance; 3. = an objective of moderate importance, 4. = an objective of major importance; 5. = the most important objective.

1. Expanding the technological knowledge base of the company.
2. Developing new technology in a product or process area.
3. Translating existing technology into new product or process designs.
4. Contributing to the improvement of existing product or process designs.
5. Developing new product or process tests to determine performance characteristics.
6. Offering new technology for reducing the cost of products or processes.

1b. Please indicate your view of the achievements of the corporate R&D lab on each of the objectives that you rated in Question 1a.

Alternatives: 1. = Very poor; 2. = Poor; 3. = Fair; 4. = Good; 5. = Very good

1. Expanding the technological knowledge base of the company.
2. Developing new technology in a product or process area.
3. Translating existing technology into new product or process designs.
4. Contributing to the improvement of existing product or process designs.
5. Developing new product or process tests to determine performance characteristics.
6. Offering new technology for reducing the cost of products or processes.

For the questions below please circle the figure (1 through 7) which best reflects your opinion.

Questions 2 through 10, 13, and 15 through 19:

Not true at all 1 2 3 4 5 6 7 Totally true

2. The R&D projects funded by the Corporate Portfolio Committee concentrate on technologies which are of great importance to the division's business.

3. The R&D projects directly funded by the divisions concentrate on technologies which are of great importance to the division's business.

4. The R&D projects funded by the Corporate Portfolio Committee align with market needs.

5. The R&D projects directly funded by the divisions align with market needs.

6. It is very difficult to get a project request incorporated into the corporate R&D lab's project portfolio.

7. It takes too long before a project can be started at the corporate R&D lab.

8. The average duration of R&D projects is too long.

9. For too many R&D projects the results are delivered later than promised.

10. The average time it takes to get answers to technical questions is too long.

11. The amount of contact with the divisions' staff during the execution of a project is.

Very unsatisfactory 1 2 3 4 5 6 7 Very satisfactory

12. The amount of contact with the division's staff after the completion of a project is:

Very unsatisfactory 1 2 3 4 5 6 7 Very satisfactory

13. The R&D staff actively seek feedback from the divisions after new products or processes have been introduced and act accordingly.

14. The reporting R&D project results is in general:

Very unclear 1 2 3 4 5 6 7 Very clear

15. Corporate R&D analyzes the complaints of the divisions in a serious manner and tries to resolve them.

16. It is important that the corporate R&D staff regularly meet with the divisions to find out what products, processes and technologies will be needed in the future.

17. It is important that there is regular staff exchange between the divisions and corporate R&D.

18. It is important that corporate R&D meets with external customers on a regular basis to assess their needs.

19. It is important that corporate R&D meets with external customers on a regular basis to assess the quality of the company's products and processes.

Appendix E. terms used in the research questionnaires

Product generation life cycle
The time-span between the moment that the first product is delivered to the external customer to the time at which the production volume is 10% of its maximum.

Operating profit margin
Operating results/revenue.

Operating results
Results after the deduction of normal operating charges and before financial expenses, taxes etc.

Revenue
Net turnover including other operating revenue, change in stocks and capitalized costs.

Revenue contribution of new products
The percentage of the current year's sales revenue derived from products and processes introduced in the last three years.

Full-time equivalents (ftes)
The extent of the appointment, i.e. a staff member with a full-time job accounts for 1.0 fte, a staff member working half-time for 0.5 fte.

Research and technology development
Research and technology development includes all basic and applied research activities directed towards expanding the knowledge base of the company. These are often directed towards the development of future projects but occur prior to specific product and/or process development activities.

Pioneering projects
Pioneering projects are R&D projects in which the product and/or process technologies are developed and implemented for the first time. These projects establish the starting point for new manufacturing processes.

Major projects
Major projects include a great deal of new product technology (more than a third of the reference system is redesigned) and/or place a significant amount of emphasis on the manufacturing processes. The amount of R&D staff input is more than 50 man-weeks.

Minor projects
Minor projects implement minor changes to existing products and processes. The scope may include a limited revision of the design to improve costs and performance. Less than a third of the design is new, and the manufacturing process is proven. R&D staff input is less than 50 man-weeks.

Concept development
In the concept development phase, the documenting and evaluating of a product concept is conducted. Information about market opportunities, competitive moves, technical possibilities and manufacturing feasibility must be combined to establish the basic architecture of the new product.

Specification and planning
In this phase, resources are typically committed to initiate the definition of the specification and development plan. This includes its conceptual design, target market, desired level of performance, investment requirements, financial impact and time-to-market. This phase ends with approval to move the project into detailed engineering.

Product and process engineering
In this phase, all aspects of the product are designed, integrated and verified. Working prototypes and the development of tools and equipment to be used in commercial production are constructed. This phase ends with an engineering 'release' or 'sign off' that signifies that the final design meets requirements.

Release to manufacturing
In this phase, the individual components built and tested on production equipment are assembled and tested as a system in the factory. During pilot production, many units of the product are produced and the ability of the new or modified manufacturing process to operate at a commercial level is tested.

Project cycle time
Project cycle time is the total time spent on an R&D project from the start of the conceptualization phase to the launch to the (internal or external) customer.

Performance to specifications
Meeting of product or process specifications.

Specification performance to customer needs
Meeting of customer needs.

Project team

A cross-functional team of scientists and engineers responsible for managing an individual project.

Networks

Networks are formal or informal regular relationships (subject-based, knowledge-based, competency-based or business process-based) transferring information outside the firm's hierarchical organization to overcome bureaucracy barriers within or between organizations.

About the author

Frances Fortuin is senior researcher at the Department of Business Administration of Wageningen University in the Netherlands where she defended her PhD thesis in 2006. She has worked as an EU expert and as consultant and researcher for a number of multinational technology-based companies. She is the vice chair of the Netherlands Eye Foundation and an advisor to the Netherlands rehabilitation fund. She has published a number of scientific articles and presented her work at international conferences in the field of innovation management. Her current research interests focus on innovation in the agrifood and in the health sector.

Printed in the United States
by Baker & Taylor Publisher Services